SMALL TOWNS AND SMALL TOWNERS

Volume 79, Sage Library of Social Research

Sage Library of Social Research

1. Caplovitz **The Merchants of Harlem**
2. Rosenau **International Studies & the Social Sciences**
3. Ashford **Ideology & Participation**
4. McGowan/Shapiro **The Comparative Study of Foreign Policy**
5. Male **The Struggle for Power**
6. Tanter **Modelling & Managing International Conflicts**
7. Catanese **Planners & Local Politics**
8. Prescott **Economic Aspects of Public Housing**
9. Parkinson **Latin America, the Cold War, & the World Powers, 1945-1973**
10. Smith **Ad Hoc Governments**
11. Gallimore et al **Culture, Behavior & Education**
12. Hallman **Neighborhood Government in a Metropolitan Setting**
13. Gelles **The Violent Home**
14. Weaver **Conflict & Control in Health Care Administration**
15. Schweigler **National Consciousness in Divided Germany**
16. Carey **Sociology & Public Affairs**
17. Lehman **Coordinating Health Care**
18. Bell/Price **The First Term**
19. Alderfer/Brown **Learning from Changing**
20. Wells/Marwell **Self-Esteem**
21. Robins **Political Institutionalization & the Integration of Elites**
22. Schonfeld **Obedience & Revolt**
23. McCready/Greeley **The Ultimate Values of the American Population**
24. Nye **Role Structure & Analysis of the Family**
25. Wehr/Washburn **Peace & World Order Systems**
26. Stewart **Children in Distress**
27. Dedring **Recent Advances in Peace & Conflict Research**
28. Czudnowski **Comparing Political Behavior**
29. Douglas **Investigative Social Research**
30. Stohl **War & Domestic Political Violence**
31. Williamson **Sons or Daughters**
32. Levi **Law & Politics in the International Society**
33. Altheide **Creating Reality**
34. Lerner **The Politics of Decision-Making**
35. Converse **The Dynamics of Party Support**
36. Newman/Price **Jails & Drug Treatment**
37. Abercrombie **The Military Chaplain**
38. Gottdiener **Planned Sprawl**
39. Lineberry **Equality & Urban Policy**
40. Morgan **Deterrence**
41. Lefebvre **The Structure of Awareness**
42. Fontana **The Last Frontier**
43. Kemper **Migration & Adaptation**
44. Caplovitz/Sherrow **The Religious Drop-Outs**
45. Nagel/Neef **The Legal Process: Modeling the System**
46. Bucher/Stelling **Becoming Professional**
47. Hiniker **Revolutionary Ideology & Chinese Reality**
48. Herman **Jewish Identity**
49. Marsh **Protest & Political Consciousness**
50. LaRossa **Conflict & Power in Marriage**
51. Abrahamsson **Bureaucracy or Participation**
52. Parkinson **The Philosophy of International Relations**
53. Lerup **Building the Unfinished**
54. Smith **Churchill's German Army**
55. Corden **Planned Cities**
56. Hallman **Small & Large Together**
57. Inciardi et al **Historical Approaches to Crime**
58. Levitan/Alderman **Warriors at Work**
59. Zurcher **The Mutable Self**
60. Teune/Mlinar **The Developmental Logic of Social Systems**
61. Garson **Group Theories of Politics**
62. Medcalf **Law & Identity**
63. Danziger **Making Budgets**
64. Damrell **Search for Identity**
65. Stotland et al **Empathy, Fantasy & Helping**
66. Aronson **Money & Power**
67. Wice **Criminal Lawyers**
68. Hoole **Evaluation Research & Development Activities**
69. Singelmann **From Agriculture to Services**
70. Seward **The American Family**
71. McCleary **Dangerous Men**
72. Nagel/Neef **Policy Analysis: In Social Science Research**
73. Rejai/Phillips **Leaders of Revolution**
74. Inbar **Routine Decision-Making**
75. Galaskiewicz **Exchange Networks & Community Politics**
76. Alkin/Daillak/White **Using Evaluations**
77. Sensat **Habermas & Marxism**
78. Matthews **The Social World of Old Women**
79. Swanson/Cohen/Swanson **Small Towns & Small Towners**

Small Towns and Small Towners:
A Framework for Survival and Growth

Bert E. Swanson, Richard A. Cohen, and Edith P. Swanson

Foreword by Harold S. Williams

Volume 79
SAGE LIBRARY OF
SOCIAL RESEARCH

 Beverly Hills   London

Copyright © 1979 by Sage Publications, Inc.

All rights reserved. No part of this book may be reproduced or utilized in any form or by any means, electronic or mechanical, including photocopying, recording, or by any information storage and retrieval system, without permission in writing from the publisher.

For information address:

SAGE PUBLICATIONS, INC.
275 South Beverly Drive
Beverly Hills, California 90212

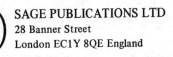

SAGE PUBLICATIONS LTD
28 Banner Street
London EC1Y 8QE England

Printed in the United States of America

Library of Congress Cataloging in Publication Data

Swanson, Bert E.
   Small towns and small towners.

   (Sage library of social research ; v. 79)
   Includes bibliographical references and index. 1. Cities and towns--United States. 2. Community. 3. Community development--United States. 4. Community leadership. 5. Regional planning--United States.
I. Cohen, Richard A., joint author. II. Swanson, Edith, 1926- joint author. III. Title.
HT123.S843   301.35'0973   78-17697
ISBN 0-8039-1017-7
ISBN 0-8039-1018-5 pbk.

FIRST PRINTING

CONTENTS

Chapter   *Page*

Acknowledgments   9

Foreword by Harold S. Williams   11

1. THE CONCERN FOR SMALL COMMUNITIES   17

Life or Death of Small Towns   17
Small Town Problems   21
The State of Knowledge About Small Towns and the Concern for Community Development   25
Contributions Toward Community Renewal   29
Community Analysis: An Aid to Renewal   35
The Focus and Format of the Book   38

2. CULTURE, NORMS, AND VALUES IN SMALL TOWNS   43

Approaches to the Study of Values, Opinion, and Beliefs   46
Small Town—Big City Value Differences   48
Variations on a Theme   51
Value Institutions   56
Values and Attitudes Toward Change   60
Conclusion   64
DISCUSSION GUIDE   67

3. SMALL TOWN SOCIAL STRUCTURE   81

Social Differentiations	82
Small Town Differentiation	85
Spatial Aspects of Social Structure	90
Socioeconomic and Political Consequences of Social Differentiation	95
Social Structure and Community Change	101
Conclusion	109
DISCUSSION GUIDE	111

4. THE ECONOMY OF THE SMALL TOWN   121

Local Economies	123
Economic Analyses	125
Economic Development	134
Conclusion	143
DISCUSSION GUIDE	146

5. PATTERNS OF INFLUENCE AND DECISION-MAKING IN SMALL TOWNS   155

Decision-Making Centers	157
Citizen Participation	161
Patterns of Leadership and Influence	164
Community Resources and Their Mobilization	170
Public Functions and Finance	173
Conclusion	178
DISCUSSION GUIDE	180

6. PROFILING COALSTREAM: COMMUNITY ANALYSIS OF A SMALL TOWN   191

The Economy	192

The Social Structure	199
Influence and Decision-Making	212
Cultural Values and Norms	226

7. COMMUNITY ANALYSIS AND CONSIDERATIONS FOR DEVELOPMENT — 237

Considerations for Community Development	241
Conclusion	261

Glossary	267
Index	275
About the Authors	279

TO OUR FATHERS

    LARS SWANSON
    MILTON COHEN
    EUGENE PINKUS

whose aspirations motivated us to work for social and political change

ACKNOWLEDGMENTS

Our appreciation goes to:

Harold S. Williams, President of the Institute on Man and Science, cognizant of the importance of joining theory and practice, set the task of reviewing the existing knowledge on small communities. He formulated a position paper which identified eight paradigms for review: (1) land use (2) social class and groups, (3) shared values and common ties, (4) distribution of power and influence, (5) functions and institutions, (6) the "Great Change," (7) social system, and (8) noncommunity. He needed to know what concepts and insights were available to him as he actively engaged in prototyping small town renewal projects.

The Department of Health, Education, and Welfare, The United States Office of Education provided some of the funds to carry out this project. However, the content does not necessarily reflect the position or policy of that agency, and no official endorsement of this manuscript should be inferred.

The field staff of the Institute on Man and Science: Perry Gordon, John LaRocca, Steve Pholar, and Jane Schautz, who shared with us their experiences.

Those who assisted with the logistics of producing the materials and manuscript: Amy Corton, Genie Corton, David Grey, Bill Hornick, Adele Koehler, Jeannette Landrum, Mary McCulloch, Patricia Reichert, and Richard Schneider.

Many "small towners," too numerous to mention in Black Lick, Corbett, Dunbar, Stump Creek, and Whitsett who generously gave and shared their hospitality, feelings, and insights with us.

Of course, none of the above are responsible for the statements herein.

Richard A. Cohen
Hoboken, New Jersey

Bert E. Swanson and
Edith P. Swanson
Gainesville, Florida

FOREWORD

This book hopes to provide a useful perspective on the small town in America. Our intent is to step back from the "firing line" of issues and problems which arise in small towns to explore some of the basic conditions which describe and even explain these communities. One basic element is physical: the buildings and patterns of land use, for example. Another is economic: trade, employment, degree of local industrial base, and distribution of wealth. A third element is governmental: the patterns by which decisions get made and carried out. And a fourth is social: including groups and organizations, shared values, and human relationships.

These basic conditions are an important framework for action. They *must* be understood in that they have such strong potential to assist or to slack action. Just as important, they are subjects for possible change in their own right. A new water system may be no more important to a town than a new system for making decisions. And fixing up walls of the fire hall may be no more important to many residents than repairing the vitality of local institutions.

The underlying framework also has great influence on the *way* things happen. And the process of dealing with problems and issues, of course, can have just as great effect on residents' happiness as the final decision itself.

In looking at the relationship between the basic elements which define small towns and the actions which residents take to live within them, our Institute is expressing one important premise: the need to better connect the worlds of theory and of practice. In improving small towns—as in so many fields—there is a large gap between those who study problems and those who most directly affect them.

On the one hand we have scholars, students, and other "thinkers" who try to understand what makes the small community "work" as it does. People in this group often use models and abstract representations to better understand what many small towns might have in common. Usually, those who study towns enter from the outside, gather data, and leave to publish in professional journals whose audience is fellow students of similar problems. Seldom do findings trickle back down to those who are studied. When they do, it is rarely with the key question addressed, "Assuming these findings are true, what should they lead you to think and to do?"

On the other hand, we have a number of "doers" who work on small town problems: some live there—local officials and leaders, volunteers, critics; others make policies and programs, at county, state, and national levels, which are equally important in changing small town life. The general reaction from members of this group is that academic studies are "impractical." Rarely do community activists seek information which should enlighten the steps they take or make more explicit the assumptions they hold.

This book is based on the premise that study and action can best be seen as two sides of the *same* coin. Studies which are divorced from any mechanism for using them and action programs that go forth in ignorance of vital information are equally deficient.

Our experience, at the Institute on Man and Science, in revitalizing small towns has led us to formulate a number of propositions about small communities. More broadly, we have drawn these conclusions which helped to clarify the need for and to shape the content of this publication.

(1) Small towns are not all alike. The generalization, "You've seen one small town, you've seen them all" is nonsense. One reason that is true is that small communities are so clearly influenced by the distinct personalities of the handfull of people most active within them. Another is the tremendous variability of small towns in terms of economic base, history, type of resident, form of governance, and so on. Small towns are complex, not simple.

(2) Small towns do tend to differ in some key respects from cities and suburbs. There are often opposing views offered on this subject by small town residents and researchers. The former often tend to feel that small towns remain set apart and distinct—little

havens from the fast pace and impersonality of the world beyond. The latter see this belief as myth, not reality, and often believe that small town characteristics have now been submerged by cultural universals.

The view reflected in this book strikes a middle ground. Most small towns are not as autonomous or even as isolated as many small towners actually feel. Yet we have seen too many distinctions to find wisdom in the belief that the small town is just one place among many where the mass society is practiced. The tables reproduced herein suggest important kinds of differences between towns and the urbanizing nation as a whole. In this respect, small town residents who want to reflect on the position of their community would do well to compare it not only to national averages but to the total picture for small towns.

(3) The concepts used to both study and assist small towns are often urban in nature and do not fit well small town realities. The urban-planning concepts have created models and approaches virtually unrecognizable by small town residents. And from the many community studies examined, we have a strong suspicion that the urbanized eye of the beholder became a filter, not a clear looking-glass.

The use of urban concepts to understand small town life is no more a problem, of course, than the use of the small town to understand cities. Thus, the assumption that an urban neighborhood carries on small town traditions as a small social unit has proven, in the instances we have seen close up, quite misleading.

(4) While small towns are distinctive places and while they do differ in certain respects from cities and suburbs, *they are far from immune from key patterns of human behavior.* Thus, people in small towns sort themselves out in terms of a status system or "pecking order." They have ways of including and excluding people. And their decision-making tends to be characterized by certain patterns.

In addition to these and other propositions about small towns, Institute staff working on community renewal problems have also evolved several interrelated theories that help to account for our enthusiasm for inquiries this book makes. The first theory develops the idea that small towns have every right to attempt to take hold of their own future—staking out directions and goals and working hard to meet them. This idea flies directly against much prevailing wisdom. Many who work with small towns feel that the job of the community is to understand the larger forces that affect it, and to learn how to adapt to them with minimal disruption. Thus, if developers enter an area, the small town should think about reusing its downtown area for parks or housing, given the inevitability of shopping centers at the edge of town. If the population of the area is growing, they too should prepare for planned growth. If it is receding, the community should brace itself for retrenchment.

In a similar, if more subtle vein, many small towns are advised to join in regionalism, gaining strength in numbers. This can be very useful strategy, especially if the region is seen as a loose federation of small social units. But often, the concept of "region" is used not to preserve small towns but to eliminate them in favor of larger sizes and scales. Regional sewage treatment, consolidated schools, planned annexations—all have a habit of destroying the integrity of the small towns within them.

Why should small towns be encouraged to plan for the future on their own terms? Because, we think, they are vitally important to America's future. This is not to say that small towns should reclaim the countryside. Indeed, nostalgia to the contrary, they never fully owned it. Rather, small towns have a place as an *alternative* to cities and suburbs. Several recent trends suggest that more and more people feel this way. One is the recently discovered tendency for people to move to small towns—especially if they are not too far from big cities. Equally important and more broad in scope is the increasingly widespread disillusionment with bigness in American life. More and more people are finding that large size—in terms not only of people but of money, power, technology, and the like—often does not solve problems but rather makes them worse. Time after time, large cities have failed to demonstrate the promised economies of large scales. And with isolated but increasing frequency some exponents of smallness—in business, communities, institutions, technologies, and so on—are showing the potential for impressive economies of very small scales.

We do not mean to say that "smallness" is *always* beautiful. Depending on its fit to function, it can be beautiful or ugly. But under certain conditions, there can assuredly be great value in personal contact, accountability, understandability, self-reliance, and the sheer elegance of things which are small. In an era we now call "postindustrial," the discovery of this truth seems surely to gather force.

If small is not inevitably beautiful, neither is the small town always charming. Indeed, as this book points out, many of the platitudes about small town residents as caring, supportive, and "close-knit" are often inaccurate. But if they are myths, at least they have some capacity to sustain. More important, they have the virtue of goal-setting. As long as we do not blind ourselves to the difference between hope and reality, the aspirations which small towns offer us yield a geography of the mind. And if we understand it well enough, we can eventually build it on our landscape.

Our hope is that you will read and react, offering your insights and criticism. We are hoping that this volume will help to initiate a new journey toward understanding the small town in America, one that includes scholars, public officials, and those who live in small towns. All of us at the Institute have a great stake in that voyage. It is a trip well worth the making.

> Harold S. Williams
> *President*
> *The Institute on Man and Science*
> *Rensselaerville, New York*

Chapter 1

THE CONCERN FOR SMALL COMMUNITIES

Life or Death of Small Towns

There is now an active concern for the viability of the small town and small towners. In part, this concern may be in reaction to the big city and large scale bureaucracies—private and public—that have come to dominate American life. The revitalization of small communities is part of the general search for smallness, simple (appropriate) technology, alternative lifestyles, and environmental conservation.[1] In community settings, smallness advocates believe that people can do many things for themselves which are presently (and often badly) done by city or regional authorities, whether that setting be the small town or urban neighborhood. Smallness advocates focus on the individual; "There is no such thing as the viability of people: people, actual persons like you and me, are viable when they can stand on their own feet and earn their keep."[2] To many, the best place for this self-reliance is not in the midst of the bustle of the city but in the relatively uncomplicated environment of the small town.

Writers for many years announced the decline of small towns and the "death" or "disappearance" of smaller villages. Simon and Gagnon, for example, wrote:

> The land and the economy of the United States will not support as many small towns as they did before. It is very difficult not to see the future as a long-drawn-out struggle for community survival, lasting for half a century, in which some battles may be won but the war will be lost. A future in which most such towns will become isolated or decayed, in which the local amenities must deteriorate, and in which there will finally be left only the aged, the inept, the very young—and the local power elite.[3]

Fuguitt, however, has found that there have been more new incorporations of nonmetropolitan towns (6,802) than dropouts (1,243) from 1900 to 1970. His studies show that: "The predominant long run tendency in changing size distribution of nonmetropolitan cities and towns is one of stability and growth. The small town or village is not disappearing."[4]

Although many small towns have declined in size in the most recent decade, there has been growth in small towns near metropolitan centers. Small places in more remote locations also were likely to grow if the county they were in contained cities with populations over 10,000.

Perhaps what is being noticed most is the loss in the number of consumer business establishments, which have declined one-third from 1950 to 1970 in nonmetropolitan towns with less than 2,500 people. This has occurred while these same places increased by one-ninth in population. Fuguitt and Beale conclude that:

> residential functions of smaller nonmetro towns are seen to have taken a contrary overall course from their business functions. Business decline does not preclude population growth in an era where there are more retired people and a greater propensity to live in one place and work in another.[5]

They believe that small town population change is "affected by size of town, location with respect to other towns, regional location, annexation policies, highway developments, and a variety of economic and social factors."[6]

As for post-1970 trends, Fuguitt and Beale note that nonmetropolitan places with under 10,000 people grew 4.9 percent from 1970 to 1973 compared to only 2.5 percent for places with over 10,000 people. Thus, there is a trend of decentralization and dispersal of the nonmetropolitan population, "smaller places and open country areas, as a class, show a revival of population increase whether they are satellitic to the larger towns or basically independent of them."[7]

The line which distinguishes small towns from big places is fuzzy and somewhat arbitrary. There is no well-defined cutoff point of population size above which a small town is no longer a small town. The Bureau of the Census provides two population categories that can be used to indicate small towns. One is the group of all communities outside of urbanized areas[8] with populations 2,500 to 10,000. The second is the settlement smaller than 2,500 which is titled "rural nonfarm." Fuguitt and Johansen divide the rural nonfarm category into two more discrete units: "villages," those incorporated places with populations below 2,500, and unincorporated areas of "nonvillage, nonfarm." What escapes these census categories are unincorporated communities such as "patch" towns, small company towns, and crossroad hamlets. Hamlets, for example, function much like villages, but are generally parts of the larger rural townships and possess no government of their own.

For our purposes, a small town is any community with a population less than 10,000.[9] Whether it is an incorporated municipality or village or an unincorporated settlement, any place that fits this basic characteristic is included in our discussion of small towns and small towners. Communities under 10,000 in population constitute the vast majority (88.9 percent) of urban and rural places in this country, but only 22.3 percent of the population live in such places (see Table 1.1).

The migration to and from small towns reflects the dynamics of citizen preferences. Post-World War II surveys "suggest that the preference for smaller scale living environments is pervasive and deeply rooted"[10] The Population Commission, for example, found in 1972 that the preference for smallness is especially strong among those residing in rural or small town areas, as 88 percent of those surveyed prefer to live in rural or small towns. They also

Table 1.1: Number of Urban and Rural Places by Population Size — 1970

Population Size	Places Number	Population Number (thousands)	%
Total	20,769	144,647	
100,000 or more	156	56,464	39.0
50,000 – 99,999	240	16,724	11.6
25,000 – 49,999	520	17,848	12.3
10,000 – 24,999	1,385	21,415	14.8
5,000 – 9,999	1,839	12,924	8.9
2,500 – 4,999	2,295	8,038	5.6
2,500 or under	14,333	11,235	7.8

SOURCE: *U.S. Statistical Abstract*, U.S. Bureau of the Census, 1974.

found that people with the following characteristics are more likely to prefer small towns:[11] (1) lower income, (2) less educated, (3) older aged, (4) Northeasterners and Southerners, and (5) whites. This preference seems to be not simply a matter of nostalgic yearning for the simpler times associated with being close to nature. Watts and Free found that:

> If left to their own devices, large numbers of young people would move out of the cities, not to the suburbs, but to towns, villages, and especially rural areas where considerably more than four out of ten said they would most like to live.[12]

The Gallup poll in 1977 found that 36 percent of those living in cities with populations of 50,000 or more would move if they had a chance, compared to only 15 percent of those living in smaller places. Only 16 percent of the central city residents who wanted to leave their city said that they preferred the suburbs—they wanted "a complete change of scenery."[13] Those who preferred to move were "the younger, better educated, more affluent, predominantly working population."[14] The key factors influencing a desire to leave urban places were high crime rates, overcrowding, poor housing, unemployment or low pay, pollution, dirt, traffic congestion, racial problems, and poor climate.

Small Town Problems

Small towns also have their problems. A 1973 conference on the Future of the Small Town concluded that small communities "were generally unable to support new residents whether with city services, health care and cultural attractions, or more fundamentally, with jobs."[15] A detailed examination of thirty-six small places with populations under 50,000 led to the identification of the following list of community needs:[16]

(1) strengthened code enforcement inspection
(2) supervised recreation programs
(3) improved fiscal capacity
(4) greater effort to fully utilize available revenue sources from non-property taxes
(5) strengthened planning process and schedule implementation
(6) more realistic review of industrial development potential
(7) increased vocational training and on-the-job training
(8) greater risk capital for businesses and support for housing market
(9) improved intracity transportation
(10) renovated central business district
(11) more adequate housing, especially low income
(12) extended public facilities such as city hall, police facilities, streets, refuse disposal facilities, sewerage treatment, and storm sewers
(13) improved health and medical services
(14) improved recreation and cultural programs
(15) improved leadership and community involvement
(16) improved intergovernmental arrangements and assistance

Small towners themselves are concerned, like people all over the country, with violence, crime, inflation, drug addiction, pollution, consumer protection, and other problems common throughout the nation. Differences between small towners and big city folks on these larger issues tend to occur in the way they each *define* and *respond to* these problems. Some of the more general issues facing small towns are urbanization, in-migration, low income, and governance.

Urbanization. Towns as small as 2,500 and less are showing signs of "urbanism." Urbanization is not just a matter of the growth of towns, though that is undeniably important, but also a change in the lifestyles of towns. Small towners are now more

Table 1.2: Differential Income Patterns by Type of Setting—1970 [17]

	Metropolitan Total	Small Towns (nonmetropolitan)		
		Communities 2,500–10,000	Village	Nonvillage Nonfarm
Median Family Income	$10,474	$8,487	$7,418	$7,477
Median Income of Unrelated Individuals	$ 2,902	$1,829	$1,741	$1,704
% of All Families Below the Poverty Level	8%	12%	16%	12%
% of Unrelated Individuals Below Poverty Level	32%	47%	47%	46%

likely to be employed in factories or offices than they are on the farms. Even those who live in rural settings and crossroads hamlets with as few as ten houses watch TV, listen to the radio, and read newspapers. This exposure to urban life tends to integrate them into a less rural lifestyle and belief system. Some aspects of urbanization are fundamentally changing the small town as we once knew it.

Lower income. Small towners earn substantially less than those who live in metropolitan areas (see Table 1.2). Perhaps this is not as serious as some analysts indicate, as small towners may desire and purchase less, but their per unit cost for whatever they purchase, a car, for example, is about the same, if not more. Considerably greater proportions of small towners, whether they be members of families or unrelated individuals, are in poverty. Lower incomes result from the smaller proportion of the labor force that is employed in white-collar occupations (see Table 1.3). Lower incomes, in turn, have resulted in housing conditions that are less adequate. This is especially true for those in poverty (see Table 1.4).

Municipal governance. Incorporated small towns are not today the self-sufficient local systems they once were or were believed to be. With rising costs for traditional governmental functions, small towns are discovering with frequency an inability to meet increasing demands for public services. In addition, other governmental

Table 1.3: Occupational Status of Employed Persons as Percent of Total Employment (1970)*

	Metropolitan Total	Small Towns (nonmetropolitan)		
		Communities 2,500–10,000	Village	Nonvillage Nonfarm
White Collar	52.2%	44.3%	39.6%	33.0%
Professional, Technical	16.0	13.8	11.8	10.0
Managers	8.5	9.0	9.6	7.0
Sales	7.6	6.8	5.7	4.7
Clerical	20.1	14.7	12.5	11.3
Blue Collar	46.7	46.6	56.5	61.6
Craftsmen, Foremen	13.6	14.3	14.8	17.5
Operatives	16.3	14.8	21.2	25.5
Laborers	4.2	5.0	5.5	6.5
Service Workers	12.6	12.5	15.0	12.1

*Does not add up to 100 percent as all farm-related individuals have been excluded.

levels are placing more demands on small towns by mandating expenditures and enforcing regulations that the communities would not have otherwise undertaken. Furthermore, small towns often do not have available the indigenous talents and expertise needed to understand and manage the variety of programs and regulations confronting them.

Influential national organizations such as the Committee for Economic Development have recommended the elimination of very small towns: "Most—if not all—of the 11,000 non-metropolitan villages with fewer than 2,500 residents should disincorporate to permit strong county governments to administer their services on a special assessment basis."[18] Small municipalities, in the name of modernization and efficiency, are threatened with losing their functional identities, sacrificing their responsibilities to more centralized units of government. The resulting loss of local control has ramifications for the very existence of small towns. In the face of increasing difficulties with small town governance, residents have a difficult choice—the easy route of shifting municipal responsibilities to higher levels, and the more difficult route of finding local solutions to small town problems.

Table 1.4: Housing Conditions (1970)

	National	Small Towns	
		Communities 2,500–10,000	Rural Nonfarm Communities
% of Households Lacking Some or all Plumbing Faciliies	4.6%	5.7%	11.9%
% of Poverty Households Lacking Some or all Plumbing Facilities	15.0%	17.0%	33.6%

Although small towns face a number of serious problems, they are not necessarily undesirable habitats. Small towns do offer particular lifestyles and residential environments which are clear alternatives to city living. Small towns represent a slower pace of life, a chance to know one's neighbors, a community which one can identify, and a place where a resident's contribution to the quality of life may be more readily noticed and appreciated. Examples of differing images of cities and small towns which motivate people to move to or remain in small towns are contained in Table 1.5.

Table 1.5: Contrasting Images of Small Towns and Big Cities[19]

Images of the Small Town	Images of the City
residents quite similar	residents quite diverse
slow rate of social and physical change	fast rate of social and physical change
lack of specialization	great specialization
subjective (feeling-based) relationships	contractual relationships
few organized groups	many organized groups
slow pace and low value placed on time	fast pace, high value placed on time
high value placed on ascribed characteristics (such as one's family, sex, other "givens" about people)	high value placed on achieved characteristics (such as educational and income levels)
high value placed on sense of permanence and security	high value placed on sense of risk-taking and the unexpected

The State of Knowledge About Small Towns and the Concern for Community Development

We know very little about small towns and small towners. There are four groups of participants who seem most concerned about small communities. One are the residents themselves, whether they be local leaders or rank-and-file citizens. The second are the policy makers at the county, regional, state, or national levels who make decisions which affect small communities and the life chances of those who live there. The third are the community developers, planners, and technicians brought in to solve a particular problem or to revitalize the whole town. The fourth group is composed of social scientists and writers who have studied and commented on the life and times of small communities.

We expect small towners to know most about their own hometown. They express concern about the decline of their communities, and seek ways to regain their economic viability or at least to protect their way of life. Yet their view may be parochial and in need of an exchange with those from the outside. Most small towners believe their town is unusual.[20]

> Each community casts a unique shadow. The imagery and interpretation of the shadows illustrate the value syndrome of each community and the similarities and dissimilarities recognized by people. These images may be reasonably accurate and changing: they may be based on only fragmentary and partial information. They may represent the limitations of one's experience in the community by reason of class position, length of residence, and what he sees, smells, hears, and reads. They may be the product of varying levels of cognition about the place where one lives, some focus on the self, the family, the neighborhood, the social group, the small community, the metropolis, the region, the nation, or even international events.

Most small towners are in a quandary. They see the image of their community changing adversely and they really do not know what to do to stop or reverse the change.

Policy makers have only recently begun to be concerned about small communities. National interests have tended to favor a *balanced* growth policy by region and by size of place. The U.S. Senate Subcommittee on Rural Development, however, believes that the United States has "a diverse mix of seemingly unrelated policies which are haphazardly and inadvertently creating a national (population) distribution policy."[21] They go on to say that it is critically important that policy initiatives consider the impact of external forces that could substantially alter the future context within which U.S. population distribution policy is made; for example:[22]

(1) energy crisis—involving transportation, consumption, and production patterns;
(2) food crisis—the use of land and water resources and the location of economic activity and therefore the location of population;
(3) environmental crisis—where some regions are too ecologically fragile to support massive population agglomerations which may force redistribution of population; and
(4) postindustrial values—the diminishing of the American "economic ethic" and the elevation of the "self-realization ethic and the ecological ethic."

The specific policy concerns for rural small towns center on their being able to serve such basic economic functions as trade and the provision of streets, water, sewerage, and so on. Rural small towns should also be able to provide services necessary to support vital primary industries (agriculture and extractive industries) as well as those required by research laboratories, universities, military bases, power generation, storage, and rehabilitation and care institutions. Socially the small town is seen as providing an alternative lifestyle to urbanism, a buffer against instability, a setting for social and political experimentation. To fulfill these functions, however, Coates believes the rural small town must have:[23]

(1) access to basic social and medical services;
(2) basic economic viability, including a range of job opportunities within acceptable travel time;
(3) a healthy and desirable physical environment;
(4) sufficient excess capacity to handle tourism, part-time residents, emergencies, and so on;
(5) a system of governance which can adapt to changing conditions (modernize and experiment); and

(6) a supportive culture that provides satisfying human interactions (cultural and artistic opportunities, education, participation in decision-making, diverse role-models for the young).

Some policy makers, especially in state government, e.g., in Pennsylvania, have begun to explore new roles they might play to increase their effectiveness in formulating their small town policy. The possibilities include:

(1) *funding agent,* so small towners can engage in some of the projects they need or want to carry out;
(2) *packager,* to relate small towners to a variety of funding agencies;
(3) *option-builder,* to expand the choices available to small towners beyond the standard solutions presently available;
(4) *facilitator,* to assist and guide small towners in achieving their objectives through self-help projects;
(5) *technician,* to provide technical assistance with experts available on call;
(6) *advocate,* for disadvantaged groups within small towns who need special attention and assistance.

Community developers, in turn, are responding to the small towns' deep-seated desires for self-betterment. They generally accept whatever national, state, and regional public policies exist and attempt to work with small towners on maximizing their existing resources and potential both within and outside the community. They believe in using a wide variety of principles, some of which are contradictory, as a basis for community action. An aspiration to improve the quality of life in small towns persists due to people's unmet needs, their commitment to succeed, and the examples of successful community efforts elsewhere.[24] Despite their resiliency, many community improvement campaigns come to naught. Much energy is spent and time donated, with at best uneven results. No one really knows how many community improvement programs have gone nowhere. Citizens and professionals have watched innumerable efforts stumble and stagnate, or just miss the target. Many small town betterment programs have been misadventures, well intended but short on achieved objectives and long on unwanted consequences.

We frequently turn to those who write about community life to learn more about small towns. While novelists and playwrights have influenced our way of feeling about small communities, the social sciences have given us a variety of ways to think about them. A review of community studies, the most often cited reservoir of knowledge, is based for the most part on: (1) individual case studies which cannot be generalized;[25] (2) selected topics which emphasize specific insights into certain places and their residents; and (3) larger comprehensive studies that sort out data by size of place and then focus on urban problems and dynamics. Community studies have compiled an impressive series of classifications, ecological maps, social stratification profiles, and structures of power. They also have provided insight into community change and conflict. While some of the concepts developed in community studies relate to small towns, they often are not pursued as a means of understanding small towns but rather as contrasts to urban life.

Perhaps the most persistent concern of writers, novelists, and social scientists is the penetration and exploitation of small towns. Not only do they believe small towns are experiencing a loss of community as external—regional and national—forces dominate the life chances of small towners, but sense that these external forces are powerful enough to exploit small communities. In addition, they are concerned about the exploitation by local elites—business and civic—who take advantage of the under class. The less-advantaged in the small town are seen to have few opportunities to escape or protect themselves from traditional exploitative practices.

Many of the members of groups concerned with small towns— practical and theoretical—have attempted to provide information and insight about small towns.[26] Some, like ants, build mounds of individual pieces of data. Others, like spiders, spin webs with logically consistent patterns of large societal systems connected to grand sweeps of historical changes. Still others, like bees, are feeding on the nectar of other people's experience and studies, digesting them, putting them into statements which attempt to provide understanding of communities as places and of the people who live there. None, however, have produced the tested theory necessary for effective intervention in small towns. They have offered a number of speculative propositions or generalizations contrasting the small town as a type of place and its lifestyle with urban places and urban life. They have also prepared descriptive statements that are unique to each place studied. They have put forth general principles on how community developers should proceed. But there is virtually no data base on the character of small towns. There are no quality of life indicators by which to assess need or to help in the development of policies and programs. What we know about community structure and dynamics has not been systematically investigated for small towns. For example, what do we know about the growth, decline, and stability of small towns or the dislocations of people? How can we account for the migration of people away from small towns, for those who remain, and for those who want to return or move there for the first time? What are the anticipated and unanticipated consequences of the many public policies, programs, and projects intended to redevelop small communities as places or small towners as people?

Contributions Toward Community Renewal

In a very real sense, the four sets of participants—small towners, policy makers, community developers, and social scientists—who have demonstrated their concern for small towns have seldom, if ever, joined together with their insights, concepts and resources with the idea of making small communities better places in which to live. Yet we believe that is what should happen if small town renewal is to take place. That is, just as there has been forming a partnership in working on urban problems, so too should there be one on small communities. To this end, we have attempted to join the interest of practitioners and theorists with the hope of providing a better understanding for those who choose to act, and a better application of the insights and knowledge for those who choose to study.

There are those—small towners, policy makers and community developers—who seem more concerned about improving small towns than spending much effort to understand what makes small communities "tick." Therefore, when action is undertaken, small towners want only to implement that which will specifically work in their town. Policy makers, especially at the national and state level, provide very general policies which do not take into account very well the unique or special features of any one town. Community developers attempt to taper and adjust the generalities of the policy makers to the specifics of a particular community. This takes not only time and care but unusual skills.

The main contributions of small towners to community renewal is their reality test of the feasibility and acceptability of proposed action. After all, they must pay the cost and live with the results. Any such test should require a fairly explicit assessment of the problems and potentials of the community. Unfortunately, there are very often only a handful of local residents consistently attempting to make this assessment and then often without much awareness of the range of available options for community renewal. Few small towners are well connected to a wide range of available resources and strategies for community action. In many small towns, those most willing to seek renewal are the "boosters" who easily settle upon an economic development strategy of bringing into town some new business. They do so without a thorough knowledge of what businesses outside of the town need to know, but more importantly, without an impact study on the most likely socioeconomic and political consequences to the community and its residents.

Public policy makers could, if they choose, apply considerable financial resources, provide technical assistance, and establish the necessary authority, especially that of state government, to address the problems of small communities. If they have not chosen to formulate a small town policy it may be that there is as yet no political constituency demanding one. There is already considerable competition for the scarce public dollar from articulate and powerful interests in urban America. The rural interests of the past were more concerned about the farm as an economically viable unit than the small town as a place that needed attention and help in renewing itself.

Furthermore, national and state policy makers tend to respond to the problems of small towns with an overly general view that is supposed to help every town, but with no specific town in mind that could actually benefit from the policy. One should not be surprised if the underlying assumption of a national public policy toward small communities would state, "What's good for the country is good for small towns and small towners."

Of all the participants who emphasize community action and improving small towns over understanding them, the most prolific have been the community developers. The most popular term given to renewal activities is community development, but it is referred to as community organization, action, and/or planning. In general, community developers are highly motivated to bring their expertise to the mobilization process. They rely on local initiative and leadership, often providing an opportunity to all segments of the community to participate in problem-solving processes. Their activities are considered democratic, rational, and oriented toward the accomplishment of specific tasks. Cary indicates that the process can be viewed as both radical and conservative.[27] It is radical in that it calls for greater citizen participation, creates new groupings and patterns of decision makers which challenge existing systems. It accelerates the pace of planned change and broadens the scope of citizen interest. It is conservative in that it helps to keep decision-making local, and centers on issues close at hand. Either way it is considered part of a "grass roots" democracy wherein local citizens recognize, define, and resolve their own community's needs. For example, the Biddles state:

> The new emerging methodology of community development supports the conviction that social improvement does not occur until the people involved believe that improvement is possible. The people themselves must be sufficiently convinced to take the initiative. The fact that they may be mired in apathy does not preclude their growing into the self-confidence of responsibility. As people are brought to feel a sense of community and to adopt goals that serve their growing concept of community, the conviction that they are able to contribute to social improvement seems to increase in them.[28]

Not all community developers subscribe to this emerging methodology. In fact, the Biddles found the development field to be suffering from confusion, with many differing interpretations about what meanings prevail:[29]

1) requiring visible installations such as highways, schools, hospitals, etc.
2) improving community services such as police and fire protection, library, recreation, etc.
3) raising economic opportunities and levels of income
4) encouraging cooperative working together
5) enhancing the underprivileged to put pressure on the powerful and/or ensuring the privileged position of the wealthier members of the community
6) stimulating the learning experience through education, research and task-oriented activity
7) increasing problem-solving capacities through handbooks-to-be-followed, or free-flowing decision-making processes
8) facilitating propaganda oriented to winning people over to a certain point of view.

The state of the art of community renewal has stimulated Khinduka to point out some of the biases of community developers to favor citizen involvement, consensus, localism, and gradualism. As a result he states that community developers display a latent propensity for delaying structural changes in the basic institutions. He points out some of their weaknesses:[30]

1) In attempting to balance models of economic growth, they overstress culturally and psychologically propitious preconditions.
2) Assuming attitudinal and value modifications necessarily precede behavioral or structural changes.
3) They often choose an incomplete, if not inappropriate, target group whose members are the victims of social and economic injustice, instead of choosing the principal beneficiaries among the privileged segments of the population.
4) Concentrating on a group's outmoded attitudes which are assumed to be the main obstacle to change. They do not recognize there may be legitimate reasons why the group has not taken the initiative or necessary risks in adopting new practices.
5) They assume man's activity cannot be changed without altering his values, which may result in neglecting the appropriate targets of intervention.
6) Pursuing a strategy for large-scale social change without concern for the time and rate of change.
7) Insisting on consensus or near-unanimity which produces superficial and innocuous activities rather than risk controversies over issues of

substance that affect diverse sub-groups differently.
8) They are slow to appreciate the need for modification of legal coercion.
9) They believe that citizen involvement is an important end in itself.
10) They promote an identification with and loyalty to the local community when it no longer exercises decisive control over the lives of an increasingly mobile population; no longer remains unaffected by external forces; and frustrates the desire for progressive changes.
11) They reinforce economically inefficient customs and practices which prolong the life of growth-resisting tradition.
12) They prefer a psychological rather than a socioeconomic approach to solving community problems.

From a review of small town development reports we discern four action-modes that are generally used in community renewal. In some of them, proposed action is a one-shot effort, while in others activities are undertaken sequentially or simultaneously. In some of them, values are made explicit at the outset, while in others project values are implicit, not clarified or justified. In some of them, outside experts play a prominent role, while in others local residents dominate the process.

The *categorical* approach tends to carry out one substantive project at a time. That is, problem-solving is based on the presumption that each problem may be solved in relative isolation, without regard to its interconnections with other problems. Recreation problems, housing problems, infrastructure problems all receive separate treatments, in some cases at cross purposes, while the cumulative direction and implications for the community structure and dynamics go unattended. There are, of course, strong incentives from the outside world for this piecemeal approach at the local level. Many state and federal grants-in-aid are available by specific problem area, inducing local problem solvers to think in categorical terms.

The *comprehensive planning* approach is intended to overcome the fragmented one through an overall assessment of community facilities and services. A checklist is followed to determine what are all the major problems requiring attention in any community. The necessary elements of the plan are duly noted and included. Recommendations on what ought to be done are made, but without much more than a superficial sense of what they add up to in the way of cost and/or impact on the residents. This approach totals the problems, but never sets them in a perspective of how they relate to social structures, decision-making systems, and community values and norms.

The *integrative* approach attempts to involve people in a process whereby they identify their own needs and preferred courses of action. These considerations are discussed simultaneously as part of a process of organizing, deciding on priorities, mobilizing support for the proposal, and engaging in the implementation of the project. Thus a specific problem, such as housing, sewage, or social services, may be the beginning of a deeper exploration of the community's problems. The integrative approach tries to connect problems to the social, economic, political, and value context of the community. In doing so, the solutions to housing problems may be found in the social structure or political system, for example, instead of in narrowly defined rehabilitation or construction actions which commonly emerge from the categorical approach.

The *dialogical* approach emphasizes "values clarification." It is concerned with having the local residents articulate their values "up front," to understand how they help or constrain achieving desired goals, and to decide the necessary changes in values and norms they must make. In many community improvement projects, the values being reinforced have tended to be those of the dominant persons or group in town. To avoid this, those who advocate the dialogical format of community problem-solving encourage community dialogue among a full range of residents to secure a more explicit understanding of community structure, dynamics, values and norms before engaging in specific projects. In this way, all problems and solutions are set in context, and residents can come to discern the cumulative impacts and implications for the whole town. Values and norms, as the "prime movers" of community action, come under direct scrutiny at the outset of the process, rather than further on down the pike as in the integrative approach. Basically, this is a process of community education, a community-wide dialogue in which participants "seek to know . . . reality (and the reasons that explain their own ways of acting) in order to transform it."[31] Dialogical action is a process by which people become aware of the forces acting upon the community from within and without.

Social scientists, more concerned with understanding than improving small towns, have attempted to formulate a number of concepts and methods to discover significant patterns of community. However, when Bell and Newby reviewed many of these studies, they found that there were about as many methods as there were community studies.[32] The variation can be seen in what Redfield believes are important ways of looking at the small community. It can be viewed as: (1) a whole, (2) an ecological system, (3) a social structure, (4) a typical biography, (5) a kind of person, (6) an outlook on life, (7) a history, (8) a community within communities, (9) a combination of opposites, and (10) whole and parts.[33] Bell and Newby, reviewing the hundreds of studies, found the community conceived as: (1) object of study and method, (2) microcosm, (3) empirical description, (4) normative prescription, (5) open and closed systems, (6) organizations, and (7) rural utopia.[34] The many patterns they have found are cited throughout the remainder of this book.

Community Analysis: An Aid to Renewal

With so many competing ways of looking at small towns, and without elaborating on any one approach, we have attempted to review the existing knowledge base of the social sciences as it pertains to small towns. We began our exploration as part of the effort of the Institute on Man and Science to revitalize small communities. For example, in preparing a feasibility study on revitalizing Coalstream (see Chapter 6), answers to a series of questions were sought: what is it that most fundamentally describes and explains this small town? is it a pattern of land use or of social stratification? is it a set of shared values, or a pervasive power structure? is it placement on a continuum of "great change" from rural to urban life, or a social system of interrelating elements? is Coalstream most basically characterized as a set of functions (such as education) and institutions (such as the school) to perform them? is Coalstream perhaps a noncommunity—one place among many where greater societal forces and values are expressed on a specific level?

To achieve our objective of reviewing available knowledge about American communities, we turned to the disciplines of sociology, political science, economics, anthropology, and so on. From them, we were able to put together four dimensions of small towns that we believe constitute the key elements necessary for a comprehensive community analysis:

(1) *cultural values and norms:* the values, beliefs, and attitudes that guide people's everyday actions;
(2) *social structure:* the differentiating of people, classes, and groups and the ways they interact;
(3) *local economy:* the production, consumption, and distribution of goods and services to the residents of the town; and
(4) *influence patterns:* the mechanisms of decision-making, opinion-shaping, and mobilizing that determine how things get done.

In a small town, these dimensions seem to be much more interconnected than in a big city. Those who are at the top of the social structure in terms of status and prestige are likely to wield disproportionate influence in community decision-making, to control economic resources, and to see their values permeate and dominate their town. To put it another way, the four dimensions of small towns, unlike big cities, are highly integrated and interrelated, like parts of the human body. To understand a small town, therefore, it is necessary to learn about its social structure, local economy, influence patterns, and cultural values and norms.

On the other hand, the complexity of big cities with large numbers of people and institutions seems to result in a fragmentation of functions, a division of power, roles, and responsibilities so that there is a likelihood of many disconnections between dimensions of a community. The interests and actions of residents associated with each of the dimensions may exist in a complimentary way or lead to conflict.

Our community analysis approach examines the social, economic, political, and value components of a community by taking them apart into their bits and pieces, and then reassembling them into what should be a more meaningful whole.³⁵ That is, rather than taking social status as a discrete phenomenon, we seek to connect it to the economic resources and political power facets of community life. We believe, for example, that learning about particular sets of values and community norms that exist in a community can provide insight into the progress or stagnation of many small towns. An analysis of social structure can expose the existence of tensions and cleavages which create a potential for community conflict. An economic analysis should reveal the availability of financial resources to support community improvement projects, and provide the basis by which to assess their effect upon the economy. An analysis of the leadership patterns and the use of authority and influence can provide insight into who gets what, when, and how.

However, this formulation of community analysis is not unique. As will be demonstrated in subsequent chapters, our understanding of values, social structures, economies, and patterns of influence is drawn from social scientists' small town case studies. Our approach is different in that rather than positing each dimension as a different and separate way of viewing the small town, it pulls together the parts as pieces of a larger whole and attempts to show the dynamic interrelationships of the parts.³⁶

How do the dynamics of these small town dimensions work? A study by Presthus of power structures in two small towns in New York State demonstrates the interactions of these dimensions. In both towns, the push for new hospitals was led by the social and political leaders of the communities. If they were not the same people, the economic and political dominants cooperated quite closely. Voluntary hospital drives were initiated, directed by the town's upper echelons, and required the mobilization of civic values to support such efforts. Presthus notes that leaders with

> [t]heir formal positions of power in the major communication, political, economic, and social institutions in the community, their shared values, and the concomitant lack of any organized basis about which effective opposition might rally, ensure that their definition of community problems and solutions will usually persist.³⁷

Thus, an understanding of the four dimensions of the small town helps students and small towners reach a more in-depth knowledge of the intracacies of the community. Residents are often very surprised by what they discover about their community. The information gathered from census reports and systematic observation should help people ward off rash solutions proposed in ignorance of the less than obvious dimensions of a small town. By studying the small town in this fashion, and describing the workings of the social structure, local economy, patterns of influence, and cultural values and norms, residents, students, and decision makers alike can get a clearer notion of both the choices faced and the possible ramifications of each decision. Only by digging into the structure of the town can one see the real-life options and impacts unfold, challenging previously held ideas about the implications of various actions. A better understanding of these dimensions also reveals more ideas on what might be done, ideas that might not be readily apparent to a "seat-of-the-pants" analysis.

Given that no one solution is the perfect solution, problems are inevitably going to arise in the future. By identifying and describing community dimensions, residents and students are better able to watch the impacts of decisions and monitor new problems as they occur. What this does is to help small towners keep pace with community problems by recognizing them when they begin to appear on the scene, not after they have already caused disruption and disagreement. An understanding of these dimensions helps small towners catch problems before they get out of hand.

The Focus and Format of the Book

We believe small towns are necessary to the survival of America. They not only played a prominent role in the development of the nation, but provide the near-infinite variety that nurtures the diversity and novelty that is so much a part of the American tradition. Our nation was born when our biggest cities—Philadelphia and New York—were about 40,000 in population. In the twentieth century we have tended to emphasize our big cities and urban problems, losing sight of the development of our small communities and the people who live there.

The focus of this book is on small towns and small towners. If they are to be viable, however, we believe that there should be a better blend of knowledge and action in their renewal. To act without understanding is inexcusable, given the insights we have developed over the years. To know without applying that understanding to the solution of community problems is equally inexcusable when needs are so great. Of course, it is not easy to bring knowledge and action together for most practitioners and theorists prefer to do their own thing. We believe that each perspective can contribute to the other, and in the process resolve some of the problems facing small towners.

In this book we have attempted to review and operationalize the existing knowledge base, so that knowledge about the community becomes an integral part of the practitioner's repertoire.

The four basic elements of community analysis are presented in Chapters 2 through 5. Chapter 2 develops the concepts of community *values and norms* and the main sources that generate them. Chapter 3 presents the *social structure,* the indicators of status and prestige and the dynamic interrelationships. The *local economy—* where small towners are employed, what they produce, and how the community benefits from business activity—is the topic of Chapter 4. The *patterns of decision-making and influence* in small towns is discussed in Chapter 5.

Each chapter begins with a discussion of the basic concepts in each of the four dimensions found in social science literature. We explore the way these concepts operate in general and in specific communities. To apply one's understanding of a dimension, we provide in each chapter a discussion guide to assist the reader in collecting information and analyzing its meaning in a selected town. The information may be readily available statistics or a recorded observation that may shed some light on a subject for discussion and interpretation.

Proceeding through the discussion guides should facilitate a way to describe and understand each aspect of a town. In other words, the setting for community renewal and the solving of community problems should be enhanced by carefully working one's way through Chapters 2 through 5. We have illustrated this process in

Chapter 6 by applying each discussion guide to a particular community where we worked for a year.

Not all small towns are actively engaged in problem-solving and community renewal. On the contrary, most undoubtedly exist from day to day without going much beyond minimum maintenance. However, most small towners and the nation will pay the price if small towns are ignored or allowed to decline. A more comprehensive and critical approach to community renewal is needed to get at the contradictions, paradoxes, and dilemmas facing small towners. Though a universally applicable answer does not exist, Chapter 7 refines our approach and a set of important considerations for those engaged in community renewal.

NOTES

1. See Harold S. Williams, "Smallness and the Small Town," *Small Town* (Vol. 8, No. 4, October 1977), p. 12.

2. E. F. Schumacher, *Small is Beautiful: Economics As If People Mattered* (New York: Harper & Row, 1973), p. 67.

3. See for example, William Simon and John H. Gagnon, "The Decline and Fall of the Small Town," *Trans-action* (Vol. 4, No. 5, April 1967), p. 51.

4. Glenn V. Fuguitt, *The Growth and Decline of Nonmetropolitan Cities and Villages* (Madison: Center for Demography and Ecology, University of Wisconsin, May 1976), p. 5.

5. Glenn V. Fuguitt and Calvin Beale, *Population Change in Nonmetropolitan Cities and Towns* (Washington, D.C.: Economic Research Service, U.S. Department of Agriculture, Agriculture Economic Report No. 323), p. 1.

6. Ibid., p. 1.

7. Ibid., p. 15.

8. Urbanized areas are generally synonymous with Standard Metropolitan Statistical Areas (SMSAs), characterized by the presence of a central (or twin cities) city of 50,000 or more and surrounding territory.

9. Jack Gibbs and Kingsley Davis also use the 10,000 person threshold to distinguish these communities from metropolitan America. In Dennis Poplin, *Communities* (New York: Macmillan, 1972), p. 41.

10. U.S. Senate Committee on Agriculture and Forestry, *City Size and the Quality of Life* (Washington, D.C.: Government Printing Office, 1975), p. 18.

11. Commission on Population Growth and the American Future, Vol. 5, *Population, Distribution, and Policy*, Part 5 (Washington, D.C.: Government Printing Office, 1972), p. 605.

12. William Watts and Lloyd Free (editors), *State of the Nation* (New York: Universe Books, 1973), p. 83.

13. Gallup Poll reported in *New York Times,* March 2, 1978, p. A14.
14. Ibid., p. A14.
15. New York Times, "Dying American Small Towns Showing Vitality and Popularity," November 6, 1973, p. C24.
16. U.S. Senate Committee on Agriculture and Forestry, *Small Community Needs* (Washington, D.C.: Government Printing Office, June 1970), pp. 5-6.
17. Data for Tables 1.2-4 have been drawn from U.S. Bureau of the Census, *Social and Economic Characteristics: United States Summary,* and Glenn V. Fuguitt and Harley F. Johansen, *The Social Characteristics of Villages in the United States* (Madison: Center for Demography and Ecology, University of Wisconsin, April 1975).
18. Committee for Economic Development, *Modernizing Local Government* (New York: Committee for Economic Development, July, 1966), p. 42.
19. Harold S. Williams, Bert E. Swanson, and Kenneth Linton, *A Community Profile: The revitalization of Stump Creek* (Rensselaerville, N.Y.: Institute on Man and Science, March 1975), p. 26.
20. Bert E. Swanson, *The Concern for Community in Urban America* (Indianapolis, Ind.: Odyssey Press, 1970), p. 115-117.
21. U.S. Senate Committee on Agriculture and Forestry, *City Size and the Quality of Life* (Washington, D.C.: Government Printing Office, June 1975), p. 15.
22. Ibid., p. 15.
23. Vary T. Coates, *Revitalization of Small Communities* (Washington, D.C.: U.S. Department of Transportation, May, 1974), p. 5-15.
24. Community developers have created a number of self-study guides, for example see Richard W. Poston, *You and Your Community* (Birmingham, Ala.: Southern University Press, 1968), and Roland L. Warren, *Studying Your Community* (New York: Russell Sage Foundation, 1965).
25. A number of early case studies are Arthur Vidich and Joseph Bensman, *Small Town in Mass Society* (Princeton, N.J.: Princeton University Press, 1968); Floyd Warner, *Democracy in Jonesville* (New York: Harper & Row, 1949); and August B. Hollingshead, *Elmtown's Youth* (New York: John Wiley, 1949). See also *Elmtown Revisited* (New York: John Wiley, 1975).
26. Bert E. Swanson, "Theories About Small Towns—Places, People and Interventions," paper presented to the American Association for the Advancement of Science, January 1974, New York City.
27. Lee J. Cary, *Community Development as a Process* (Columbia, Mo.: University of Missouri Press, 1970), p. 5.
28. William W. Biddle and Loureide J. Biddle, *The Community Development Process* (New York: Holt, Rinehart and Winston, 1966), p. viii.
29. Ibid., pp. 281-296.
30. S. K. Khinduka, "Community Development: Potentials and Limitations," *Readings in Community Organization Practice,* edited by Ralph M. Kramer and Harry Specht (Englewood Cliffs, N.J.: Prentice-Hall, 1975), Second Edition, pp. 175-183.
31. Paulo Freire, "Education as Cultural Action—An Introduction," paper presented to the 1970 CICOP Conference, March 1970, p. 5.
32. Colin Bell and Howard Newby, *Community Studies* (New York: Praeger, 1973), p. 54.
33. Robert Redfield, *The Little Community* (Chicago: University of Chicago Press, 1955).
34. Bell and Newby, op. cit. See Chapter 1 and 2, pp. 21-81.

35. See Figure 7.1 in Chapter 7 for a schematic diagram of the four overlapping and overlaying concepts.

36. See, especially for a speculative linkage, Bert E. Swanson and Edith Swanson, *Discovering the Community* (New York: Irvington Publishers, 1977), pp. 387-388.

37. Robert Presthus, *Men at the Top* (New York: Oxford University Press, 1964), p. 404.

Chapter 2

CULTURE, NORMS, AND VALUES IN SMALL TOWNS

Americans are known as a very pragmatic people with the operating rule of, "If it works, it's OK." As a nation we seldom spell out our goals or any guiding principles by which to achieve them. As de Tocqueville pointed out early in our history:

> Americans are always in action, and each of their actions absorbs their faculties: the zeal which they display in business puts out the enthusiasm they might otherwise entertain for ideas ... it is extremely difficult to excite the enthusiasm of a democratic people for any theory which has not a palpable, direct, and immediate connection with the daily occupations of life.[1]

Yet few other nations have made explicit the set of philosophic goals that provides legal guidelines for the conduct of human affairs as set forth in the Preamble to the Constitution. The five key goals reflecting our prevailing values are to: (1) establish justice, (2) insure domestic tranquility, (3) provide for the common defense, (4) promote the general welfare, and (5) secure the blessings of liberty.[2] The values enunciated in the Constitution permeate all governmental action, and therefore serve as the guidelines for what ought and what ought not to be done by national, state, and local government officials.

But even nongovernmental action seems to be guided by a set of operational precepts. Lipset has examined American history and discussed the paradoxical relationship of two essential values—equality and achievement. He believes the two are not entirely compatible as each engenders a reaction which threatens the other. Equality, for example, requires that all persons must be given respect because they are humans, regardless of their high or low status. The contending value of achievement means that an individual must assume responsibility to gain the necessary capacity and skills for success. Lipset describes the dynamic:

> For people to be equal, they need a chance to become equal. Success, therefore, should be attainable by all, no matter what the accidents of birth, class, or race. Achievement is a function of equality of opportunity. That this emphasis on achievement must lead to new inequalities of status and to the use of corrupt means to secure and maintain high position is the ever recreated and renewed American dilemma.[3]

Thus, Americans do have a set of well-established guiding principles. However, they are not absolute nor consensually held for all time and places. Each region of the country and each generation seems to distinguish itself from others. In fact, the mosaic of American values, with centers of differing cultural norms, is believed to be a basic element of this country's "democratic" way of life.

The American culture and its norms are, unfortunately, more assumed than studied. There are innumerable lists compiled by those who have attempted to provide an inventory of some basic American values:[4]

(1) *Freedom*—the power to determine one's own action.
(2) *Individualism*—a right to be different, to rely upon one's own abilities, and to cherish one's specific beliefs.
(3) *Democracy*—a system combining majority rule and protection of minority rights.
(4) *Practicality*—the adaptation of something for actual use.
(5) *Pecuniary evaluation*—the determining of worth by money or materialistic things.

(6) *Success*—the gaining of wealth and position or achieving one's objectives.
(7) *Education*—the acquisition of knowledge, training, and skills.
(8) *Science*—a systematic knowledge of the physical or material world.
(9) *Happiness*—a belief that life should be enjoyable and that pleasure is better than pain.
(10) *Humanitarianism*—a belief that man is obligated to be concerned with the welfare of the human race as a whole and as individuals.
(11) *Conformity*—a general agreement with others and compliance with those in authority.

While these values may be generally accepted by most Americans, it is important to note that some are seemingly contradictory such as individualism and conformity. Some values are more widely accepted in particular settings than in others. Perhaps most noticeable is the difference in what is valued from one town to another (San Francisco—elegance, Chicago—virility).[5]

As one might expect, American values seem to differ from those of other nations. In a value study of community leaders in India, Poland, United States, and Yugoslavia, Jacobs and his colleagues found that American leaders rejected the values of conflict avoidance and economic equality. They also have the lowest commitment to selflessness and economic development. American leaders were less committed to innovative change, more action-oriented, and placed great value on citizen participation in decision-making. They were no different than their counterparts in other nations in their commitment to honesty and to their country.

> Thus, the rejection of social restraints and the assertion of his own freedom of action seem to constitute the core of the American leader's value orientation.... [R]eliance on expertise or authority figures is rejected, perhaps because it reduces the leader's freedom of action. The leader's identification with local as opposed to national interests seems to free him from restraints that may operate on him from an external and distant source (the nation). The relatively low commitment to economic equality may be similarly interpreted; while the first involves long-term personal sacrifices, the latter implies a commitment to removal of economic disparities. Both are social commitments that limit the individual's freedom of action. The relatively lower stress on Selflessness seems also to conform to this interpretation, as it implies less willingness to subordinate personal self-interests to that of the community or society.[6]

Approaches to the Study of Values, Opinions, and Beliefs

Social science tends to use and measure a number of terms to refer to a person's preference for one side or another on a community decision, issue, controversy, or conflict. The term *value,* for example, has come to mean such things as anything appreciated, anything desired, any object of need, desirable end states which guide social action, a normative standard which influences human choice among alternative courses of action.

Opinions are topical short-run judgments or impressions on questions of public affairs. *Attitudes* are more enduring and comprehensive views or connections. *Beliefs* are more fundamental sentiments on the central values of life. Opinions and attitudes are outgrowths of beliefs and are more amenable to change. Again, many of these terms are used interchangeably, a reflection on the underdevelopment of this important topic.[7] Nevertheless one should try to distinguish between the terms, especially if one notes that a person expresses one opinion on a subject matter but when it comes time to carry it out, behaves quite unexpectedly. The importance of discerning the differences between opinions and behavior has been expressed by Jacobs:

> Consciously or unconsciously, people conceal the actual grounds for their actions in a cloak of widely approved "values" (the process often called rationalization) or attempt to manipulate the action of others by persuading them that a given action either fulfills worthy ends, or contrarily, violates cherished principles. But the very tendency to seek justification, in terms of standards that others would deem acceptable, is itself a recognition of a deep-seated, widespread disposition to apply values in the process of making choices and determining conduct.[8]

Social scientists are interested in measuring the prevailing value patterns, the direction of attitudes, and the intensity of feelings held by various population segments of the community. They also have been interested in understanding the determinants of attitudes and their change. They have developed a number of ways to measure and monitor community values: (1) archive records, (2) public opinion polls, and (3) participant observation. These are not offered here in order of preference but should be used when appropriate given certain resources and skills to carry them out.

The amount of information available in the written record on small towns is surprising. For example, at city hall can be found election results that represent the political preferences of the voters. Similarly, local budgets reflect the priorities of public expenditures of elected officials and their constituents. Even the police "blotter" and court docket indicate what is not permissible behavior in town. Past copies of the newspaper should also reflect what is important in town. However, because the paper is privately owned and operated, one might expect that it more aptly represents the views and values of the publisher and/or his major advertisers. Webb, Campbell, Schwartz, and Sechrest encourage the use of the less continuous or episodic and private record which appears in sales records, institutional records, and personal letters and diaries. They believe that even though there are risks in relying on someone else's selective filter, the Chinese proverb prevails: "The palest ink is clearer than the best memory."[9]

Public opinion polls have become the most popular means of measuring the opinions, attitudes, and beliefs of a sample of citizens. While few surveys are conducted on particular communities, national surveys are carried out regularly. Therefore, to learn what small towners think, one can turn to these national surveys. Often, Gallup polls report the answers of those who live in different sized communities.[10] Pollsters generally explore several areas of public concern. One is the determination of voters' preferences for candidates in forthcoming elections. A second is a measurement of citizen loyalty to certain groups, such as political parties, trade associations, labor unions, and so on. A third is the identification of citizen preferences on specific public policies and issues such as medical care, housing, crime, and the like. A fourth is an analysis of the broader concern of evaluating society and its workings in general. The results of such surveys indicate the direction in which citizens believe or prefer their community to move, measure the extent of consensus or agreement on the course of events, and assess the intensity or depth of citizen support either for or against what is happening or what is proposed for the future. Some analysts use survey findings to predict particular elections, while others, more cautious perhaps, use them as guides for public policy. Interviews with community leaders—because they are involved in the process—sometimes enhance our understanding of the value implications of particular decisions.

Participant observation is used to enhance understanding of the relevant values underlying community affairs. Here the *participant-as-observer* and the *observer-as-participant* lie somewhere on a continuum between the *complete participant* (the emphasis of a resident) on one end and the *complete observer* (the emphasis of a pure researcher) on the other.[11] A community resident, directly affected by specific public policies, could if he or she desired, become analytic and better understand what is happening in the community and what could or should be done differently. The purpose of the researcher's involvement is to enhance insight into the innermeaning of community action and to uncover the value dynamic. Either role contains the hazards of being distracted from one's normal pursuit by reducing the resident's involvement or increasing the researcher's identification with those with whom he works.

Hornick and Enk have suggested some important points about values and human behavior:[12]

(1) Neither beliefs, attitudes, nor behavior can be directly observed or measured; they must be inferred from a person's statements and/or behavior.
(2) One needs to determine the difference between reported and observed values and those implicit in behavior.
(3) No person's or group's value system is fixed; the hierarchy of values changes—often temporarily.
(4) Values serve a person by orienting him in a complex world and by helping him to act—especially by helping him to act quickly in complex situations.
(5) Values vary in space and time as a result of varying needs, environment, socialization and social structure. Rapid change in any one creates stress for change in the others.
(6) Values can be demonstrated to be sound or unsound in terms of the consequences they produce.

Small Town—Big City Value Differences

In what way do the values and attitudes of small towners differ from big city folk? We shall explore the value studies of small towns to determine whether values in small towns differ from those in big cities, and/or whether there are major differences between various groups and individuals within small towns. A

1970 Harris survey in the states of Washington and Wisconsin found that relative to the past, 22 percent of the residents of cities compared to 34 percent of small town residents thought their community was getting better than it had been a few years before. Nearly half the city inhabitants, but only 19 percent of the town residents, thought their community was *not* good. Relative to the future (the next five years), only 29 percent of the city residents expected their community to be better compared to 34 percent who expected it to get worse. In contrast, 38 percent of the residents of towns expected their communities to get better and 18 percent expected them to get worse.[13] One might interpret these differing expectations as a measure of small town optimism based on past improvements. These expressions are, of course, the residents' subjective sense of what is happening to conditions in their own communities. These differences are further reinforced by the evaluation residents of varying-sized places give to a number of community attributes. Overall, big city dwellers "tend to be considerably less satisfied with their communities than are residents of suburbs, small cities, and especially of rural areas."[14]

In race relations, Harris found more city folks preferred that blacks be integrated into the white society. He also found that fewer small towners believe blacks are justified in their demands for social change, and feel that blacks are moving too fast to improve their lot.[15] This might be interpreted as small towners being less tolerant toward racial minorities, and less supportive of their efforts to improve.

Perhaps a better reflector of citizen values can be found in the priorities they establish in government spending. Watts and Free believe that, "Deviations from the norm in spending patterns among segments of the population are, of course, indicative of different value systems at work within these groups."[16] Table 2.1 indicates the relative government spending priorities of those who live in cities, suburbs, towns or villages, and rural areas. Small towners agree with city dwellers that there should be increased spending for crime prevention, drug addicts, and water pollution. City residents are more prone to support increased spending for senior citizens, medical care and Medicaid, air pollution, education for low income children, and higher education. They both agree to current levels of expenditures for urban problems, parks and recreation. City dwellers are more inclined to keep present spend-

Table 2.1: Government Spending Priorities (Points) — 1972

Item	National Totals	City	Suburbs	Town or Village	Rural
Crime	88	89	87	88	92
Senior Citizens	87	91	87	83	86
Drugs and Addicts	86	85	85	82	93
Water Pollution	81	82	85	70	75
Air Pollution	80	80	85	76	77
Educate low income child	80	83	81	79	77
Medical Care	80	84	77	77	77
Medicaid	74	80	71	76	66
Higher Education	72	75	72	76	68
Urban Renewal	69	72	70	68	58
Urban Problems	68	69	72	69	60
Jobs for Unemployed	67	69	65	64	68
Public Parks	66	66	72	64	62
Mass Transportation	66	70	66	65	58
Housing in General	64	68	68	59	57
Highways	63	62	60	63	66
Public Housing	62	70	59	62	54
Black Americans	57	66	54	58	45
Welfare	53	60	48	55	45
Space Program	25	23	28	26	26

SOURCE: Adapted from Watts and Free, pp. 320-322.

ing levels for jobs for the unemployed, mass transportation, housing in general and public housing. Small towners, however, are less inclined than city folks to support current levels of spending for black Americans and welfare, but more inclined than those living in suburbs and rural areas.

There is considerable disagreement among social scientists on the question of whether ideologies operate at all in America.[17] Some have questioned whether specific value patterns are more likely to prevail in small towns than in larger cities. Ladd found relatively little or no ideological basis for community decision-making either in Hartford, Connecticut (population 155,000), a large central city, or Putnam (population 8,400), located at the outskirts of Hartford. In fact, he believes the small town is contemptuous of ideology because it experiences a politics "of narrow issues, limited government activity, immediate self-interest, and personal foibles."[18] Other analysts question whether there should be any community values operating to form and support attachments to some locality other than the nation. Lane, for example, believes the deep sense of community of small towns: (1) stimulates enduring loyalties to a place, (2) creates immobility, (3) increases attachment to conventional approaches, (4) enforces a narrow moral code of behavior, (5) personalizes issues and events, (6) generates myths and heroes, and (7) ensures for solidarity among the citizenry.[19] All these recognized virtues of the small town threaten the larger democracy and the ability of the nation to survive. He therefore proposes to remove the sense of community as a layer between the individual and the nation.

Variations on a Theme

In an overview of urban places, Swanson concluded, "that each community casts a unique shadow. The imagery and interpretation of the shadow illustrate the value syndrome of each community and the similarities and dissimilarities recognized by people."[20]

Unfortunately, there are few comparative community studies that document the common as well as unique nature of American communities, especially of small town value systems. Two studies of small towns did, however, find considerable value pattern differences.

The Institute on Man and Science, while engaged in a revitalizing effort in Stump Creek, Pennsylvania, a small (former) mining community, compared the beliefs and values of two nearby small towns. One was Helvetia, a former mining town, and the other was Charlestown Village, a low income rental housing project. The contrasting value patterns are summarized for each town.

Stump Creek.[21] Perhaps the most widespread value expressed in Stump Creek is faith in the Lord, and to a lesser extent in the self. This faith is closely bound to church doctrine, particularly for the 50 percent of the residents who are Catholic. Religious traditions influence the strong sense of family, expressed values such as honesty and the Golden Rule, and belief in the essential "rightness" of traditional roles and beliefs. Just as important, it provides a justification for the way things are in the world—a platform for acceptance.

The ethic of hard work and of getting along on whatever the fruits of one's labor bring in is very important in most Stump Creek households. So, too, is the capacity for endurance. Residents take pride in their ability to cope with their conditions, relatively low incomes, and negligent landlords. Special pride is taken by some long-term residents in responding successfully to conditions of increasing physical deterioration.

Many residents seem to have willingly sacrificed the benefits that higher incomes and a more modern physical setting might bring in return for a perceived sense of independence, which takes form as an avoidance of challenge and of complexity. Residents have not generally sought "better" jobs, for example, that would have them do the bidding of relatively few more highly placed persons. Instead, they prefer the roles they have and seem to cope very competently—with the comfort that no one will bother them because they do not make waves. Competition is reserved for the baseball field, where it is taken very seriously. Also, many wage earners are doing what they had hoped to do. Their expectations have been met.

The value of independence from the outside world is tempered by the importance placed on a feeling of interdependence within Stump Creek. Residents are expected to keep their own house in order in a way that will be acceptable to others and to fit the pattern of Stump Creek. While people often have limited contact, there is an awareness that when problems arise, you can count on the help of the community.

Values in Stump Creek are a composite of external influences. From Eastern European cultures come the sense that man's role must be strong and that loss of job is emasculating. From the mines come respect for authority and the power of the external system in it. From religion comes a foundation for accepting adversity while not giving in to it. And from the local schools come a recognition of the necessity for order and discipline, and the concept that education, above all else, is the key to improvement.

More recent lessons are not ignored, however. It was an abrupt transition from a company-owned mining town to a town with a landlord interested in rental profits. The resulting decline led to an intensive distrust in institutions, including both the government and private sector. There is a way in which social institutions are not working in their interests—and as the community has continued to decline, the lesson becomes more ingrained.

What is distinctive in Stump Creek is not the set of influences which are active there, but rather the way in which, over the years, residents have translated, mixed, and modified those influences to bring about something quite uniquely Stump Creek. The family is the major mediating force in this process, selecting and filtering—from mother to daughter, father to son.

There is a strong belief that Stump Creek is different—and better—than most other places. Much of this feeling is expressed as a shared dislike for cities—and the high crime, impersonality, and complexity attributed to them.

An important explanation for this shared belief lies in the selective migrational pattern when the mine closed down. The descriptions given by residents of those who stayed and those who left suggest that the characteristic of risk-taking is important. Those who did not want to face uncertainty in the world beyond elected to remain in Stump Creek.

It is certainly true that such factors as television, schools, the selective out-migration of former residents, in-migration of more dependent populations, and other factors of the mass society are influencing traditional values, especially among the young. But change—or more accurately, response to change—is very slow in this relatively isolated, geographically distinct small community. On one hand, children go to school and absorb the message that formal education is the key to a better life, while on the other, they quietly persist in their modest career goals, recognizing that more education is needed to seek the same job as one's father.

Helvetia.[22] A sense of asceticism, piety, and devotion to hard work pervades Helvetia. Many residents are proud of their thrift, and quite rigorous in bringing their wants and means into balance. Some are compulsive about remaining debt-free, regardless of how low their income.

Many current practices in Helvetia are traceable to the Slavic origins of the miners, and to community activities prominent during the mining days. Catholics, for example, still separate on Sunday, the Lithuanians going to one town to worship and the Poles to another. Pride in ethnicity remains. It seems to reflect the fact that people were dealt out of the American system at birth and never had access to it later.

As in Charlestown, there is a strong premium placed on noninvolvement and privacy. The implications of these values in Helvetia are quite different, however. The outcome of privacy in Helvetia is quietness, taken to mean the lack of interaction as well as of noise. There is a very limited amount of daily visiting by families and individuals. The natural setting appears also to be prized. "Nature" seems to connote the absence of pollution, crime, and other urban ills more than any positive values of its own.

Also in contrast to Charlestown, there is a great deal of respect for authority. This seems connected with a quite positive value placed on passivity and faith in a "welfare state," run first by the mine and now by a real estate family partnership. In Charlestown, people generally do not complain because it does not bring positive results. In Helvetia, people do not complain because complaining is not considered legitimate. There is a predominant sense of submissiveness. Few residents report any participation in working on a community problem.

Charlestown Village.[23] The changing nature of the project has precluded—and probably will continue to preclude—the development of many stable norms and values. A "hard work" ethic does not prevail, nor do related religious values. While there are frequent clashes among residents over what "ought to be," no interest is especially well organized, with the exception of several individuals who take personal steps to back up their viewpoints about drinking and noisiness. Further, most clashes seem based far more on the situation than on abiding values.

The value that is probably most deeply held is that of privacy, and the implicit right not to be bothered or even governed. This seems to have encouraged the atmosphere in which common exteriors cover extraordinary differences in life-style. For a few, distinctiveness is expressed in the highly personalized sanctuaries they have made of their living space. For others, the apartment seems only a shell to encompass disheveled artifacts of daily life—still another aspect of living over which they seem to have lost control.

Many residents voice great concern with material possessions, and indicate that money is an overriding factor in personal happiness. Even the students living here say that earning money is the goal of their education and the way to escape from their dependent status. Many residents do not seem to relate the concept of money to a detailed sense of the goods and services it can buy, but rather to the perceived general position or status it could potentially afford them.

Although *all* residents of Charlestown receive public assistance, there is little evidence of the welfare syndrome. Individuals do not seem to be part of generational welfare families. There is little sense of embarrassment at being on assistance, and a sense that their dependency is not total.

Presthus, in comparing two upstate New York communities not only discovered considerably different value patterns in the two town, but found an important difference between the beliefs of community leaders and those of the rank and file citizens. On the whole, Riverview exhibited more "liberal" values and the rank-and-file members of both communities were more "pro-government." The upper-class citizens of Edgewood were somewhat lower on alienation. Authoritarianism was characteristic of both communities and significantly higher among the lower socioeconomic status persons. In explaining the greater degree of tension and conflict between Riverview's political and economic leaders, Presthus found:

> a significantly higher proportion of members of Riverview business and political organizations rank high on authoritarianism. Such ideological differences and attending differential rates of participation found in the two communities probably rests to some extent upon differences in their social structure.... [R]egardless of class differences, Edgewood citizens are more committed to participation.[24]

These two communities contrast sharply with the impression that small towns are consensual where a relatively homogeneous population agrees on basic values. However, even if the community is highly stratified there may be considerable agreement as those in the upper strata may control and shape the prevailing values. Thus, when Vidich and Bensman found equality, industriousness, improvement, and optimism to be the dominant ideologies in Springdale, they also discovered that certain values were emphasized by those in different positions within the social stratification of the community. For example, in classifying the disposition toward income, they found that prosperous farmers preferred *investments*; merchants preferred *savings*; professionals preferred *work* and *consumption*; while "shack people" were *consumption*-oriented. They contend therefore that:

there is a plurality of alternative bases for class divisions and life stylizations in the rural community. Moreover, these life stylizations have their locus in different and specific segments of the community. The extent to which any one perspective predominates in the community at large depends on the relative success of the representatives of each of the ideologies ... the underlying secular trend indicates a shift from production to consumption values in the community.[25]

Value Institutions

Value patterns can develop locally in the autonomous community or reflect externally induced values in the penetrated place. There also can be an admixture of local, regional, and national values. Much speculation has occurred toward identifying the major institutional generators of value systems.

Five major institutions essential to almost any society, are generally found to play significant roles in small towns: (1) schools, (2) families, (3) religious institutions, (4) economic firms, and (5) governmental units. While these institutions serve many functions in community life, we discuss them here primarily as generating, perpetuating, nurturing, and protecting those values, beliefs, attitudes, and opinions preferred by the influential and used to guide their own organizational members and others in town.

Where there is a value of consensus in a small community it is generally associated with a homogeneous population. It may also be due to the existence of little diversity in the types of families, religious congregations, alternative schools, competing economic firms, or strong ideologically based political parties and/or leadership. It is expected that all these institutions agree on what patterns of values should prevail. Disagreement, of course, is likely to develop if these institutions work at cross-purposes.

Public School. Schools, generally supervised by state agencies and run by consolidated school boards (representing an area larger than the immediate community), are sometimes the source of dissension in small towns. Representatives from different towns bring to the school board different perspectives and beliefs, and a struggle for control over competing values ensues. A major and most perplexing issue in small towns is the purpose or orientation of local education. School graduates leave the small town and seldom return. Thus, as Martindale and Hanson state, the question small towners face concerning their children's education is

> whether it should be primarily directed to the needs of the local community or to the success of young people in the world at large. To direct education to the first may enhance the local community at the expense of those who migrate; to direct education to the second requirement may be to build in a mechanism that tends to pump the finest intelligence and talent of local young people into the world at large.[26]

A related question is whether students should be trained with occupational skills or given a more general and moral education that integrates the individual into the larger societal order. Parochial and private schools are sometimes established, although less likely in small towns, to meet alternative educational preferences. Other educational disputes occur over increasing taxes, the discipline of students, the social conduct of teachers, and prayer in the schools. Thus, the consensus of small towns is likely to decrease as diversity within a major institution occurs.

Family. Similarly, as different types of families develop or locate in a town, a diversity of life-styles develops; a challenge to the predominant "nuclear" (two adult) family is presented by an "attenuated" (one parent) family. The former, threatened by an alternative life-style, perceives the latter negatively, especially if it is a female who heads the household and requires public assistance or Aid to Families with Dependent Children (AFDC). An issue develops over the virtue of self-sufficiency as opposed to dependency; in the process much abuse is heaped upon those labeled as "reliefers." Arensberg and Kimball identify the following values as important to families: "privacy of the home, sanctity of womanhood, fulfillment of obligations (especially those based on one's word), rights of property, defense of one's honor, and resistance to the intruder."[27]

Some community disputes occur over differences in child-rearing practices. Those often frowned upon are general permissiveness, disorderly behavior, and the acceptance of unwed mothers. Yet, even though such family differences can be the focus of community concern, very little attention has been given to the family in the study of community decision-making. This may be due to a lack of research on understanding the role of the family in community affairs or recognition that the family is more likely to resist change than facilitate it.

Religion. Most towns have a variety of religious groups indicating different religious values. There is probably less variety in small towns as many are predominantly Protestant. However, there is considerable divergence among various Protestant sects. The introduction of an additional religious congregation in small communities often can bring about competition between diverse doctrines. This competition is more than a matter of institutional rivalry. Each religion has its own doctrine which affects how its respective members perceive community life and which determines their particular values. For example, Weber formulated a major distinction between the Catholics' and the Protestants' orientations. Catholics assume the answers to life's problems are known, and that those who are wise and good will accept and obey the precepts derived from them. Protestants, by contrast, assume that all life's important problems are not yet known but they can be discovered by rational means. Lenski, confirming Weber's hypothesis, indicates the importance religious values play upon community affairs by saying,

> What is possible, what is probable, and what is inevitable in any given secular organization is a function, in part, of the characteristics of the individuals who staff them; and this in turn is a function, in part, of the socio-religious groups to which they belong.[28]

Many community studies tend to overlook the importance of religion in community affairs. Pope provided one possible explanation for this absence when he noted that ministers and churches did not become involved in community issues nor censure unjust economic conditions. He states that, "Confusion, individualism, moralism, and evangelism are characteristic of their social outlook and nullify the actual application of a professed desire and ability to create a better world."[29] Yet most communities experience issues that are based on religious values. These not only involve the obvious religious issues such as "release time" for religious instructions, the enforcement of "blue laws" on Sunday prohibiting commercial activity and consumption of alcoholic beverages, but also the reform of local government, desegregation of public institutions, and the fluoridation of the water supply.

Economy. Those who determine money matters play a significant role in community affairs. Form and Miller believe that "There is probably no area of life—government, education, religion, welfare, and family—which is unaffected by the decisions made by these economic agencies."[30] A one-company town is known for its general dominance of community life. An introduction of diversity into economic structures, discussed in Chapter 4, can result from a number of developments—a shift from dominance by one company to the presence of several competing firms, the replacement of a locally owned firm by a subsidiary or affiliate of an absentee-owned corporation, a change in the community's economic function from residential to industrial, a major change in the relative prosperity of the local economy, or a shift of local services from the private to the public sector (or vice versa). These structural changes can introduce an element of dissension in the previously harmonious community. When changes threaten the small town's continued "smallness," the value of growth may come under attack.

Shifts in the relative proportion of services provided by the public sector may stimulate challenges to the commitment of some economic dominants toward maintaining a minimal scope of government. Economic downturns, manifested in changes in the occupational structure, can even raise questions about the reliance on private entrepreneurial activity for economic livelihoods. The questions resulting from changes in economic institutions are likely to be:

(1) how much faith do we put in continued growth?
(2) how much should the public sector be involved in providing what has previously been provided by the private sector?
(3) how much faith do we place in the responsibility and stability of private business?
(4) how will change in the economic institutions affect changes in other institutions?

Government. Government is generally perceived as the one institution to which *all* residents are members and from which specific rights and responsibilities flow. It is the one institution wherein all citizens are supposed to stand equal before the law, and all are to have equal access to those who make decisions for the community as a whole. We have come to expect differences of opinion on what should be done and refer to this interplay of differences as politics. Political consensus is likely to be disturbed by changes in the formal decision-making arrangement and/or in the informal patterns of power and influences discussed in Chapter 5. That is, governmental authority is generally divided by various units of local government—municipal, county, education districts, and other special districts—and especially newly created regional authorities.

Political controversy develops over differences on the purposes, costs and benefits, and policy-making procedures of local government. Similarly, as influence patterns shift from consensual to competitive due to policy differences, the struggle may lead beyond competition to confrontation or conflict. Differences are most frequently expressed over what is the proper scope of government. That is, some advocate the expansion of local government functions to include more and more activities and programs such as urban renewal, economic development, or providing parks and recreation. Others oppose these proposals as activities which will simply increase local taxes or as activities best performed by private enterprise.

Values and Attitudes Toward Change

We have demonstrated that each community has some kind of prevailing value pattern that is generated and nurtured by primary institutions. We have also indicated that small town values are somewhat different from those formed in big cities. Further we have shown some distinctions between the values of leaders and the rank-and-file. We have discussed the diversity in preferences and ways of life between various groups and classes within the small town. One should also expect that proposed change, whether it be locally initiated or externally induced, will be affected by and in turn will affect, the prevailing value pattern. Each community will modulate its agenda of change through filters of values, culture norms, opinions, and attitudes held by individuals, groups, and the community.

Warren, in discussing the "great changes," has noted a number of major value changes in American communities:[31]

(1) The gradual acceptance of governmental activity as a positive value in an increasing number of fields.
(2) The gradual change from a moral to a causal interpretation of human behavior.
(3) The change in community approach to social problems from that of moral reform to that of planning.
(4) The change of emphasis from work and production to enjoyment and consumption.

No one has established the determinants of these changes nor the public policy and community development consequences. Elazar, studying migratory patterns in American history, has discerned three major political subcultures which mold perceptions about the role of government, influence those who are active in public affairs, and direct the way local government is used. The first category, termed *individualistic*,

> emphasizes the centrality of private concerns; it places a premium on limiting community intervention—whether governmental or nongovernmental—into private activities to the minimum necessary to keep the market place in proper working order.[32]

The individualistic subculture has an overriding commitment to the marketplace and will not initiate new programs unless demanded by public opinion. It is ambivalent toward bureaucracies. Cities in the Middle West fall into this group. The second, *moralistic* subculture, emphasizes the "commonwealth" and good government as one of the greatest activities of man. There is

> a general commitment of utilizing communal—preferably nongovernmental, but governmental if necessary—power to intervene into the sphere of "private" activities when it is considered necessary to do so for the public good or the well-being of the community, utilizing governmental resources for social regulation as well as for the promotion of economic well-being.[33]

This culture is a product of Puritan New England and the more secularized "Yankee" influences. The third, *traditionalistic*, accepts a hierarchical society with those on top, an elite, playing a dominant role in government. Government, in turn, takes a positive but limited role in "securing the continued maintenance of the existing social order."[34] This political subculture is mainly associated with the plantation agrarianism of the South. Some communities, of course, may contain any combination of these three subcultures.

Another framework for understanding the dynamics of local decision-making has been developed by Agger, Goldrich, and Swanson. They examined how communities decide upon expanding, maintaining, or contracting the local scope of government. One major dispute generally experienced in most communities from time to time is over "who shall rule?" Agger et al. found a variety of answers to this important question: (1) industrialists-financiers, (2) proprietors-professionals, (3) public officials-administrators, (4) the propertied, (5) disadvantaged businessmen, (6) labor minorities, and (7) racists—black or white. They associated some of the following ideologies with these answers: (1) orthodox conservatives, (2) progressive conservatives, (3) community conservationists, (4) Jefferson conservatives, (5) radical rightists, (6) liberals, (7) radical leftists, and (8) racial supremacists—black and white.[35]

Many specific disputes occur over such public policies as the need for urban renewal, water and sewer systems, police and fire protection, and many other proposals which seek to establish or redirect the functions of local government.

While it is very difficult to generalize on what values prevail in small towns, there are a few general impressions worth mentioning. Ladd found values expressed in "perennial pining" (something ought to be done) over personal and public morality, and tax spending.[36] Vidich and Bensman noted that candidates for public office (representatives) in Springdale "must subscribe to a low-tax, low-expenditure ideology and be relatively unsophisticated in the techniques of political analysis and public administration."[37] Presthus discovered both leaders and citizens preferred self-reliance and "private" initiative over government solutions to local problems.[38]

Much of the above discussion has focused on specific values, institutions, and disputes. A more general concern of leaders and citizens alike is the level of consensus (harmony) and/or conflict which generally prevails among individuals, groups, and institutions; there are varying threshold levels of conflict which small towners believe should not be surpassed. Efforts for community improvement or change might be resisted by those comfortable with keeping things the way they are. When problems persist unresolved, the demands of residents (and outsiders) for something to be done have a tendency to grow in intensity. Occasionally, these demands reach the point of developing into community conflict. At the heart of many small town controversies are value conflicts, again a reason to get to the roots of community values for effective problem-solving. Table 2.2 demonstrates the value conflicts in the design and dynamics of a new town located on the East Coast.

Table 2.2: Value Conflicts and Tradeoffs

Desired Value	Conflict or Tradeoff
(1) make Columbia a beautiful place	either maintain beauty with development, controls and limitations, or allow individuals to alter their properties as homeowners and possibly reduce natural beauty
(2) build a sense of community and belonging	either maintain a homogeneous population with communal activities, or allow more privatism and heterogeneity, possibly weakening the sense of community
(3) create a secure family-oriented environment	either focus amenities and institutions toward families or include the needs of singles and retirees, possibly diluting the family orientation of the town
(4) conscious, rational community-planning	either give control to expert planners and limit popular participation, or allow more participation and tolerate the corresponding decrease in thoroughly rational planning
(5) create a comfortable life of amenities and services	either preplan for all needs and thus create a controlled, rather unexciting lifestyle, or allow for more unplanned solutions (and unmet needs) but more options for creative problem-solving

SOURCE: Adapted from *American Values and Habitat: A Research Agenda* (American Association for the Advancement of Science, 1976, Appendix A), pp. 187–188.

Coleman offers three criteria which suggest that an event may evolve into a controversy: (1) the event must touch upon an important aspect of the community members' lives (education of one's children, means of livelihood, religion, and so on); (2) the event must affect the lives of different community members differently (a tax policy that affects homeowners one way and renters another); (3) the event must be one on which the community members feel that action can be taken and not one which leaves the community helpless.[39] Different types of towns generate conflicts of varying styles and substance. Self-contained towns, where people both live and work, generate intense conflicts in response to economic issues. Suburban or bedroom communities, where residents commute to work out of town, seem to erupt over tax issues, due to the impact tax increases have on homeowners. They also experience controversies over differing perceptions of educational values, political beliefs, and patriotic concerns. Service towns or resort communities face conflicts between the interests and behavior of long-time residents and temporary summer or winter vacationers. Finally, towns without much industry but highly politicized by governmental authority (patronage, governmental contracting, and so on) will likely face more politically charged controversies.[40]

Donahue believes the acceptable threshold of conflict in smaller towns is lower than in big cities because the small towns "tend to repress conflict and to forestall action when conflict occurs."[41]

Conclusion

We place a heavy emphasis on community values and norms, partly because they are so little studied and understood, and partly because they play a significant role in the community development process. They not only shape and mold the social, economic, and political dimensions of community, but these structures protect and preserve the prevailing community values and norms. Values tend to be overlooked by many community developers. For them it is certainly much easier to plot strategies for downtown redevelopment—straight bricks and mortar—than to try to change deep-seated attitudes and beliefs about the function of downtown. Similarly, putting in new storefront facades, home improvements, flood walls, and paved streets, while meeting immediately perceived needs, may not in themselves address some of a town's deeper value issues, such as the fairness of services, treatment of social groups, the openness of decision-making, or the overall image of the community.

Therefore, we believe that community renewal in itself is not a neutral activity. In the dynamic interrelationships of the four elements, actions taken in one sector will more than likely have side effects on other dimensions. That is, a governmental reform may relate to value changes, and so on. But some actions make the role of values central to the planned activity and community development in order to affect values, or try to change values in order to achieve progress in community renewal. The alternatives ignore values until it is too late, when dominant community norms have warped the original program intentions or resisted them altogether. In the case of conflicting interest groups, where one group is favored over another, with whom does the community developer side? If the decision of majority rule is chosen, then the action is no longer value-neutral. Any decision in a case of conflicting interest automatically calls to the forefront the values and beliefs of those making decisions on small town community improvement goals.

Blizek and Cederbloom, for example, have characterized the willingness of community developers to foster self-determination, community autonomy, and local participation as *normative neutrality*. They criticize this position of neutrality on the grounds that the developers might find themselves perpetuating some form of social injustice, such as assisting a segregated community to exclude people of other racial, ethnic, or religious backgrounds. They also believe that the developer cannot remain neutral as he will often have to choose between competing interests within a community. They believe if community development practice is to avoid unjust or self-defeating activity, then normative principles of justice must be articulated for resolving conflicts of interest within and between communities.[42] Stewart on the other hand not only accepts the designation of normative neutrality for community developers, but believes that position needs to be strengthened rather than weakened. He believes it is both presumptuous and unnecessary for community developers to formulate principles of social justice as a guideline for practice. Instead, the role of the professional is to assist the community to examine and interpret its needs; to distinguish between means and ends; to understand the possible present, near future, and long-term implications of action; to see, in this light, what the options are and the likely consequences of each... a neutral comprehensive advice that refers to the whole cultural setting, political climate, and trends in socio-political morality of the larger society of which the client community is part.[43]

The argument above is not over whether community development is value-free but whether there ought to be a set of explicit principles that will assure social justice. The Biddles assert clearly as to values: "let it be stated immediately that development is always normative. That is, it takes place in the midst of actions that serve chosen goals."[44]

DISCUSSION GUIDE

The study of a community's values and norms is a difficult task and remains relatively undeveloped in small town research. Public opinion polls are costly, difficult, and a sometimes unreliable method to learn what individuals will do based on what they profess to believe. We propose several approaches to explore what cultural norms and values prevail in a specific small town. One is to examine community decisions and issues that have occurred in recent years or may occur hypothetically in the foreseeable future. A second approach is to explore institutional roles and how they affect values. A third is to "vote" on a budget to establish what would be a community's priorities. The final approach provides a general inventory of the quality of life factors small towners believe are important.

I. Community Decisions and Issues

To better understand what opinions, attitudes, and beliefs shape community action, the following steps should be used.

A. *Identify a few important decisions and/or issues* that developed in the recent past or are about to be or should be resolved in the community. One might review the list of goods and services provided in Table 5.2. Some community decisions and issues may be to: (check relevant ones)

_____ 1. install sewer treatment, water system, and so on
_____ 2. provide housing for low-income families
_____ 3. increase local property taxes for municipality/schools
_____ 4. regulate land use, and establish housing code
_____ 5. locate industrial firm
_____ 6. attempt government reorganization and/or reform
_____ 7. provide parking and traffic control
_____ 8. fluoridate water supply
_____ 9. others (list)

B. *Key Informants and Participants.* To obtain reliable information on some of the above community decisions, one might consult knowledgeable persons in town or scan past copies of the local newspaper accounts on what happened. In Chapter 5 we indicate three approaches—positional, reputational, and decisional—to identify some of the most probable key participants in town. These may include: (name key participants)

(1) newspaper editor or reporter _____
(2) local banker or manager of an economic firm _____
(3) mayor or city councilman _____
(4) president of local civic organization _____
(5) others (list) _____

C. *Reasons for Supporting or Opposing.* To discover the operational opinions, attitudes, and beliefs one should obtain the arguments used by advocates on one or the other of community issues. These statements should be sorted by groups which seem to be generally in agreement. That is, not only assemble the statements into those which support and those which oppose a particular proposal, but sort into groups of commonly shared opinions and beliefs. Two arguments may differ on the proposal but do so by sharing the same values yet applying them differently.

ISSUE: _____

Group 1   Group 2

_____   _____
_____   _____
_____   _____
_____   _____

Group 1 statements:
_____
_____
_____

Group 2 statements:
_____
_____
_____

II. Institutional Roles and Values

It is difficult to discern whether the arguments and statements collected are those of individuals or represent groups in town. It is most important to learn what roles various local institutions took or will take on community decisions and issues. Therefore, a brief review of their expressed opinions, attitudes, and beliefs should help identify some of the important prevailing cultural norms and values that persist in town.

After having selected a decision or issue that has been or is likely to be important in the community, indicate the following for each institution:

(1) Did or would it play a role?

    a. government_____
    b. economic firms_____
    c. schools_____
    d. church_____
    e. family_____
    f. other (specify)_____

(2) What position did or would it take and for what stated or likely reasons:

    a. government_____

    b. economic firms_____

    c. schools_____

    d. church_____

    e. family_____

    f. other (specify)_____

(3) What key values seem to be reflected in their stated positions, i.e., identify values discussed in this chapter.

    a. government_____

    b. economic firms_____

c. schools_____

d. church_____

e. family_____

f. other (specify)_____

(4) What effect or influence did or would it likely have on the final outcome?

a. government_____

b. economic firms_____

c. schools_____

d. church_____

e. family_____

f. other (specify)_____

A. *Government.* As elected representatives of the people, local public officials generally are key persons reflecting the public interests on taxes, on initiation of new projects, on priorities of what public functions should be performed, and on distribution of public services. Sometimes difference of opinion over these decisions arises along political party lines, or between those in authority and their critics. The critics may comprise a competing elite organized to oppose specific policies and to offer themselves as replacements for those in office. Small towns have a general reputation for adopting and accepting a minimum scope of government, which keeps public spending and taxes low. Such a policy may reflect not only a preference for private programs over public programs, but also a concern over the cost of public goods and services for those less able to pay (e.g., those with fixed incomes). Advocates of increased governmental activity generally believe that the public or the community has an obligation to meet the needs of people. Of course, they may be simply serving their own special interest in providing for the educational and recreational needs of their own children.

Some of the most interesting value differences are revealed by determining who pays for and who receives public services. While there are few services provided in small towns, those that are provided are carefully watched. For instance, many families notice if some families receive more benefits than others, such as better garbage collection, street repairs, or police protection. Disagreements over the availability of public services can reveal whether there are those who believe in equal treatment, those who believe some should be denied services because they are lazy, unacceptable people and such action may encourage them to leave town.

Another common issue is procedural in that some persons express concern that government should ensure that the "rules of the game" and fairness be followed in community decision-making.

B. *Economic Firms.* Businessmen, largely viewed as making jobs available and providing goods and services, generally play an important role in civic affairs. Perhaps those economic firms that hire more workers are looked on as more important. The preferences of merchants and manufacturers are listened to on such policies as the growth and stability of the community, the use of public or private enterprise to provide certain services, the use of local taxes to carry out governmental functions, and the development of land-use regulations. While some merchants prefer a vital downtown, others support land-use proposals which allow for the commercial or industrial development on the outer edge of town. Both may justify their support on the grounds that their alternative will enhance growth, provide jobs, and generally benefit the community. The competition over land-use decisions should reveal the relative value placed on self-interest as compared to public interest expressed by certain economic firms and how the land can and should be used to enhance the general community. In some cases, advocates make an important distinction between the direct short-range benefits from growth in contrast to the more indirect, long-range costs of providing services to an increasing population.

C. *Schools.* Local educators, who teach the community's youth for some twelve years are expected to socialize them to the country's value system as well as prepare them to enter the marketplace, to participate in civic affairs, and to become adults. These objectives are supposed to be well integrated while reflecting the expressed opinions and beliefs of the local community. As educators defend their need for increased taxes, new building, additional curriculum opportunities such as vocational training, and the selection of particular textbooks, they may and should reveal some of the basic educational values actually being implemented. Similarly, their supporters and opponents in and out of the Parent and Teachers Association (PTA) demonstrate their beliefs as they express their opinions on what should be happening with the youth of the community. Not all these decisions are made locally; neither do the participants—school officials, parents, and other citizens—always make explicit their underlying values. However, they may reveal, for example, the degree of importance they attach to science and technology in students' education. Some very divergent views on morality and the role of religion and family surface in disputes over sex education or release time for religious instruction.

D. *Religious Institutions.* Church leaders—clergy and laymen—are looked to for a well-developed doctrine which explains man's role in the universe and which provides an explanation of what constitutes ethical relations between humans. Perhaps in a society which provides for the separation of church and state, one should not expect much involvement by religious congregations in community affairs; yet there are a number of community decisions and issues which seems to attract religious spokesmen, or wherein religious beliefs appear as the basis for position-taking on public issues. It is relatively easy for religious leaders to make clear why they demand release time for religious instruction, and where they stand on moral questions relating to family life. However, it may be much more difficult for them to see the immediate relevance of religious doctrines to civic affairs, or to clearly articulate the applicable beliefs and have them accepted. For example, to translate the doctrine of social equality of all mankind into the everyday experience of all residents has yet to be achieved, though it is a doctrine supported by religious moral codes *and* the law on equal opportunity which is so difficult to enforce. Another example, more personal, is abortion and the "right to life" and whether local health agencies should facilitate or inhibit individual decisions. Similarly, many cross-currents of forces develop over whether individuals can or should believe in progress and improve themselves and their communities, or accept their fate as part of a larger order.

E. *Family.* The family as an institution is not organized as the other institutions are and is therefore less visible. It is difficult to learn what family values or orientations exist in small towns. The PTA should be one source of information on family values, at least insofar as children and school affairs are concerned. Some communities have family planning units which are essentially interested in birth control and assisting female-headed households. Another source of information on family values would be the social caseworker who supervises those families on public assistance or relief. Very often in small towns concern is expressed, not in any formal way, about the presence of juvenile delinquents, prostitutes, and reliefers, but how these persons are perceived as a disturbance to the community or constitute a source of community problems is not clear. Other family-related problems develop over the special needs of an antipoverty program for the poor or recreational programs for youth.

III. *Choosing Community Priorities*

The third approach to identifying cultural norms and values would be to reexamine the use of community resources—public and private, monetary and volunteer. To estimate the use of resources applied toward the well-being of the residents and the town as a whole, one should conduct an inventory of community activities. This would include not only the budgets of local governments, but the efforts of the volunteer fire department and ambulance service as well as the wide variety of community programs—civic, recreational, and the like. Estimates should be calculated in dollar amounts, person-hours served, and so on. Once the present pattern of resources is roughly known, then "votes" can be taken to determine whether the present expenditure pattern of community resources is preferable or whether some new priorities would evolve. It is important to discuss the reasons for retaining the present or changing to a different pattern of priorities. Very often these discussions are difficult because many persons cannot possibly understand where new resources, notably monetary, can come from. While this approach is often "realistic," especially in small towns, and forces advocates of change and those who propose new programs or projects to face up to finding the resources, it often inhibits any further consideration of possible changes.

We suggest the following three different topics for discussion.

A. *Public choices on changing the local government's budget.* Which items of the local budget would you change and by how much (use a "+" for increase, a "0" for no change, a "-" for decrease)? Most important is to discuss why an item of expenditure is the largest.

	Per Capita		Direction of change (+,0,-)
	Present	Future	
1. General revenues	$_____	$_____	_____
2. Intergovernmental aid	_____	_____	_____
3. Property taxes	_____	_____	_____
4. Other taxes	_____	_____	_____
5. Charges/fees	_____	_____	_____
6. Debt	_____	_____	_____
7. General expenditures	_____	_____	_____
8. Highway	_____	_____	_____
9. Hospital	_____	_____	_____
10. Health	_____	_____	_____
11. Fire	_____	_____	_____
12. Police	_____	_____	_____
13. Sewerage	_____	_____	_____
14. Parks/recreation	_____	_____	_____
15. Housing/urban renewal	_____	_____	_____
16. Libraries	_____	_____	_____
17. Other (specify)	_____	_____	_____

State the items which should be increased, remain the same, or decreased, and discuss the reasons for each proferred change.

Increase: Reasons
_____   _____
_____   _____

Remain the same: Reasons:
_____   _____
_____   _____

Decrease: Reasons:
_____   _____
_____   _____

B. *Community organizational activity.* Which community activities would you change and by how much? Again indicate why.

		$	Present Person hrs.	$	Future person hrs.	Direction of change (+,0,−)
1.	Volunteer fire	__	_____	__	_____	_____
2.	Youth recreation	__	_____	__	_____	_____
3.	Beautification	__	_____	__	_____	_____
4.	Environmental concerns	__	_____	__	_____	_____
5.	Clean-ups	__	_____	__	_____	_____
6.	Festivals, parades	__	_____	__	_____	_____
7.	Art shows	__	_____	__	_____	_____
8.	Religious fellowships	__	_____	__	_____	_____
9.	Adult recreation	__	_____	__	_____	_____
10.	Other (specify)	__	_____	__	_____	_____

Increase: Reasons:
_____   _____
_____   _____

Remain the same: Reasons:
_____   _____
_____   _____
_____   _____

Decrease: Reasons:
_____   _____
_____   _____
_____   _____

State the items which should be increased, remain the same, or decreased and discuss the reasons for each preferred change.

C. *Allocating new money.* With the provision of federal General Revenue Sharing or the "block" grants program of the Housing and Community Development Act of 1974, on what programs, projects, and activities should $100,000 be spent? Again what reasons best support your preferences?

Item	$	Person hours	Reasons
1. _____	___	_____	_____
2. _____	___	_____	_____
3. _____	___	_____	_____
4. _____	___	_____	_____
5. _____	___	_____	_____
6. _____	___	_____	_____

IV. Quality of Life Preferences

The final method of ascertaining the values of residents in a community is to survey their hopes and fears. The following questions might be asked of a sample of residents or by the small group assembled to discuss community affairs.

All of us want things out of life. When you think about what really matters in your life, what are your wishes and hopes for the future? In other words, if you imagined your future in the best possible light, what would your life look like then, if you are to be happy? Take your time in answering; such things aren't easy to put into words.

_____
_____
_____
_____
_____

Compare your answers with those of the nation for 1974.

	Personal Hopes	U.S. %	Community %
1.	Better or decent standard of living	29	_____
2.	Good health for self	28	_____
3.	Aspirations for children	24	_____
4.	Peace in the world	16	_____
5.	Happy family life	15	_____
6.	Economic stability in general	15	_____
7.	Good health for family	11	_____
8.	Own a house or live in better one	11	_____
9.	Good job, congenial work	11	_____
10.	Better or more honest government	10	_____
11.	Peace of mind, environmental stability	9	_____
12.	Leisure time	8	_____
13.	Wealth	8	_____
14.	Happy old age	8	_____
15.	Better world, understanding, cooperation	8	_____
16.	Concern about or assistance to relatives	7	_____
17.	Social justice, eliminate discrimination	7	_____
18.	Employment	6	_____
19.	Safety from crime	6	_____
20.	Self-improvement	5	_____
21.	Acceptance by others	5	_____
22.	Resolution of personal, religious, ethical problems	5	_____
23.	Christian revival in general	5	_____
24.	Success in one's work	5	_____

Now, take the other side of the picture. What are your fears and worries about the future? In other words, if you imagine your future in the *worst* possible light, what would your life look like then? Again, take your time in answering.

	Personal Fears	U.S. %	Community %
1.	Economic instability	26	
2.	Ill health for self	25	
3.	War	18	
4.	Lower standard of living	16	
5.	Ill health in family	12	
6.	Unemployment	12	
7.	Inadequate opportunities	10	
8.	Crime	9	
9.	Social decay, ethical, religious	9	
10.	Lack of freedom	6	
11.	Political instability, dissent, unrest	6	
12.	Shortages of various kinds	6	
13.	Lack of integrity in politics and government	5	
14.	Drug problems	5	

V. Interpreting Community Norms and Values

From the materials collected above and closely listening to the statements and discussions on why certain decisions, issues, and priorities ought or ought not to be adopted, one can now proceed to interpret the expressed opinions, attitudes, and beliefs into meaningful patterns of community norms and values.

(1) What are some of prominent clusters of values that seem to be used in community decisions, issues, and setting priorities?
(2) Which opinions, attitudes, and beliefs seem to be most agreed upon (consensual)?
(3) Which opinions, attitudes, and beliefs seem to be most disagreed upon (competitive)?
(4) How strongly held and resistant to change are some values?

Using the list of values presented by Warren, one can explore some of the value clusters that are likely to play a role in shaping the decisional preferences of small towners. These are most likely to become evident when a controversial issue develops as it tends to reveal covert values that operate implicitly most of the time. In small towns it is expected that there will be greater consensus on values—making behavior more predictable. A consensual pattern would also indicate general agreement between and among the five major institutions. That is, there would be support and mutual reinforcement by each because a clear respect for each other's point of view would exist, or because no distinguishable difference among the preferences of the churches, the schools, local government, or the business community would exist. The greater the disagreement, the more diverse the values are likely to be. For example, conflicting values are stimulated by the effort to reduce or eliminate air and water pollution which may threaten the closing of a local industry and result in a loss of jobs.

Some disagreements may involve minor differences over where to put parking meters along Main Street. However, disagreement can develop into a major dispute; e.g., developing a viable downtown shopping area benefiting someone's interest versus an alternative site benefiting another developer. Differences may occur over *how* to meet a particular objective rather than over the objective itself. It should be noted that while families generally confine their differences to the immediate household and religious groups to their churches, local government and the schools are most likely to be centers of many open community disputes. Competition between economic firms is expected. However, they prefer a consensual community, believing it is better for business to avoid controversy which tends to discourage customers. It should also be noted that the adoption of new programs, such as urban renewal, is more likely to bring disagreement than is the modification of existing programs.

In a community with contending sets of leaders, political parties, and institutions, prediction is more difficult and less accurate. Other factors such as resources, influence, and mobilization must also be considered. If one believes community leaders accurately represent rank and file citizens, then one should explore what beliefs and attitudes the prevailing leaders hold and prefer. If the leaders are not representative, then a greater effort should be made to discuss with a broader spectrum of the public what they hold dear, what they want their community to be like, and what their policy preferences are on a specific decision. Challenges over "who shall rule" often reveal deep-seated values. Similarly, controversies are more likely to expose firmly held beliefs than are routine decisions.

Finally, in a consensual town values may seem to be firmly held, yet there may never have been a challenge to test their strength. Challenges from internal or external forces can demonstrate which sets of values are actually firmly held and which are likely to give way.

It should be clear by now that values are difficult to identify, collect, and analyze. They remain an underdeveloped aspect of social science and community development, especially as one moves from what *is* to what *ought* to be the quality of life in small towns. Therefore, it is most difficult to indicate what values will prevail in what type of community. Yet it is worth speculation when one considers whether or in what way a community has improved, or desires or intends to improve itself.

NOTES

1. Alexis De Tocqueville, *Democracy in America* (New York: New American Library, 1956), p. 271.
2. Preamble to The Constitution of The United States of America.
3. Seymour M. Lipset, *The First New Nation* (New York: Basic Books, 1963), p. 2.
4. Roland Warren, *The Community in America* (Chicago: Rand McNally, 1963), pp. 87-89.
5. Yi-Fu Tuan, *Topophilia: A Study of Environmental Perceptions,* Attitudes and Values (Englewood Cliffs, N.J.: Prentice-Hall, 1974), pp. 192-224.
6. Philip E. Jacob, *Values and the Active Community* (New York: Free Press, 1971), p. 82.
7. For these distinctions see Bernard Berelson and Gary A. Steiner, *Human Behavior* (New York: Harcourt, Brace and Jovanovich, 1964), pp. 557-558.
8. Philip E. Jacobs, "Values and Public Vitality," in *Values and the Active Community* (New York: Free Press, 1971), p. 7.
9. Eugene J. Webb, Donald T. Campbell, Richard D. Schwartz, and Lee Sechrest, *Unobtrusive Measures* (Chicago: Rand McNally, 1966), p. 111.
10. *The Gallup Opinion Survey,* Princeton (selected years).
11. Raymond L. Gold, "Roles in Sociological Field Observation," *Social Forces* 36 (March, 1958), pp. 217-223.
12. William F. Hornick and Gordon A. Enk, *Value Issues in Technology Assessment* (Rensslaerville, N.Y.: (Institute on Man and Science, 1978), p. 3, 23-25.
13. Louis Harris, *A Survey of Public Attitudes Toward Urban Problems,* November, 1970, p. 9.
14. Angus Campbell, Phillip E. Converse, and William L. Rodgers, *The Quality of American Life* (New York: Russell Sage Foundation, 1976), pp. 236-237.
15. Louis Harris, *The Harris Survey Yearbook of Public Opinion-1970* (Louis Harris & Associates, 1971), pp. 504-506.
16. William Watts and Lloyd A. Free, *State of the Nation* (New York: Universe Books, 1973), p. 253.
17. For a discussion on political ideologies see Robert E. Agger, Daniel Goodrich, and Bert E. Swanson, *The Rulers and The Ruled* (Scituate, Mass.: Duxbury, 1972), pp. 9-21.
18. Everett Carll Ladd, Jr., *Ideology in America* (Ithaca, N.Y.: Cornell University Press, 1969), p. 146.
19. Robert Lane, *Political Ideology* (New York: Free Press, 1962), pp. 226-227.
20. Bert E. Swanson, *The Concern for Community in Urban America* (Indianapolis, Ind.: Odyssey Press, 1970), p. 115.
21. Harold S. Williams, Bert E. Swanson, and Kenneth Linton, *A Community Profile: The Revitalization of Stump Creek,* Paper Number Two (Rensslaerville, N.Y.: Institute on Man and Science, 1975), pp. 24-27.
22. Ibid., pp. 62-63.

23. Ibid., p. 67.
24. Robert Presthus, *Men At the Top* (New York: Oxford University Press, 1964), pp. 366-367.
25. Arthur Vidich and Joseph Bensman, *Small Town in Mass Society* (Princeton, N.J.: Princeton University Press, 1968), p. 78.
26. Don Martindale and R. Galen Hanson, *Small Town and the Nation* (Westport, Conn.: Greenwood Press 1969), pp. 114-115.
27. Conrad M. Arensberg and Solon T. Kimball, *Culture and Community* (New York: Harcourt, Brace and Jovanovich, 1965), p. 172.
28. Gerhard Lenski, *The Religious Factor* (New York: Doubleday, 1961), pp. 343-344.
29. Liston Pope, *Millhands and Preachers* (New Haven, Conn.: Yale University Press, 1942), p. 186.
30. William H. Form and Delbert C. Miller, *Industry, Labor and Community* (New York: Harper and Row, 1960), p. 4.
31. Roland Warren, op. cit., pp. 89-84.
32. Daniel J. Elazar, *Cities of the Prairie* (New York: Basic Books, 1970), p. 260.
33. Ibid., p. 262.
34. Ibid., p. 264.
35. Agger, Goodrich and Swanson, op. cit., p. 15.
36. Ladd, op. cit., p. 190.
37. Vidich and Bensman, op. cit., p. 115.
38. Presthus, op. cit., p. 428.
39. James S. Coleman, *Community Conflict* (New York: Free Press, 1957), p. 4.
40. Ibid., pp. 6-7.
41. George Donohue, "Feasible Options for Social Action," Larry E. Whitney, ed., *Communities Left Behind* (Ames: Iowa State University Press, 1974), p. 126.
42. William L. Blizek and Jerry Cederbloom, "Community Development and Social Justice," *Journal of Community Development Society* (Vol. 4, No. 2, Fall 1973), p. 45.
43. Guy S. Stewart, "Conflict, Social Justice and Neutrality," *Journal of Community Development Society* (Vol. 5, No. 1, Spring 1974), p. 6.
44. William W. Biddle and Lourlide Biddle, *The Community Development Process* (New York: Holt, Rinehart and Winston, 1966), p. 2.

Chapter 3

SMALL TOWN SOCIAL STRUCTURE

A basic tenet of the American society is stated in the Declaration of Independence: "We hold these truths to be self-evident, that all men are created equal... that they are endowed by their creator with certain inalienable rights that among these are life, liberty, and the pursuit of happiness." Yet around us we see inequalities, the rich and the poor, those well-housed in good neighborhoods and those housed in shacks or slums, those with secure jobs and those accustomed to unemployment or public assistance. Although we all may be born equal, differences quickly emerge. In the end, we do not all partake or benefit equally in the opportunities of this country. The hard reality is that the cards are stacked against some groups and individuals. The belief in equal opportunity is not always translated into equal results. Differences in family education, income, occupational status, and other characteristics lead to social distinctions that shape the options and opportunities which each individual faces, regardless of the promise of the Declaration of Independence.

References to social inequalities appear in our everyday language. The phrases "upper class," "lower class," and most frequently "middle class" are common descriptions applied to people encountered at work and around town. To deal with the variations within categories, "middle class" is divided into the "upper-middle" and "lower-middle," and so on, reflecting the complexity of social groupings throughout American society. Other terms such as "blue collar," "working class," or "white collar" indicate social inequalities based on the division of labor. Differences in life-styles are also apparent. People notice who lives in the better neighborhoods, who owns the largest, most spacious houses, who drives the fancier, more expensive cars. For some people, the terminology "social inequality" is an abstraction, an intellectual method for sorting out and classifying people. For others, social inequality is part of the reality of life, and the language of inequality becomes a guidepost and cue for behavior.

Social Differentiations

Widespread and persistent inequalities have led social scientists to examine American communities for the factors that contribute to and the consequences that flow from social differentiations. They have attempted to measure the overall social profile, the vertical cleavage lines within a community, as well as variation in the "life chances" of individuals and groups that result from one social structure compared to another.

More specifically, Svalastoga had discerned in four different stages of economic development five patterns of social differentiation: preagrarian societies (egalitarian model); agrarian societies (caste model and estate model); early industrialization (class model), and Western industrial societies (continuous or status model).[1]

Most of the early understandings of social structure did not begin with the town as a unit of analysis, but was seen from the perspective of a whole society or nation. Many insights were drawn from European theorists, particularly Karl Marx and Max Weber. These two nineteenth century thinkers contributed some fundamental concepts that in modified form have persisted to this day. Marx's insights into the economic processes of Europe's period of industrialization and growth were accompanied by a very comprehensive analysis of social structure. The common use of the concept of *class* goes back to his definition. According to class analysis, the individual's position in the system of production and distribution of goods or commodities determines his or her position in the class structure. Everything else about the individual—his wealth, property, and other material conditions—stemmed from his location in the productive system.[2] Although Marx described several types of groups, in the end there were really only two classes: those who owned the means of production and those who worked for the owners.[3] Each group has a *consciousness* of its own—as there is a chronic division of interest between them based on *exploitation*—those who control the means of production exploit those who produce—which leads to *conflict* and a reorganization of society.

Weber on the other hand, while accepting Marx's notion of class, grounded firmly in economics, added two other noneconomic concepts: (1) "status," to indicate social honor or prestige, and (2) "party," representing social or legal power.[4] That is, class, status, and party or power combine to influence one's social behavior. A person with high status, party, and class positions would be likely to associate with people perceived to be of the same social rank. The ideas of Marx and Weber have strongly influenced the approaches to social stratification being used to understand American communities today.

While most reference to social caste has been made to conditions in India, there have been some studies of American communities in the South that reflect similar conditions based on race. That is, social selection is determined by ascription (birth), whereby: (1) membership is hereditary and fixed; (2) the choice of spouse is endogamous (kept within the caste); (3) contact with other castes is limited and restricted; (4) consciousness of membership is emphasized by name, customs, and so on; (5) the caste is united by a common occupation, religious, tribal or racial origin; (6) relative prestige of different castes is well established and jealously guarded.

Myrdal, in his study of Negroes in America, believed that many of these conditions prevailed in varying degrees by region, and between rural and urban communities as the American creed on equality of opportunity is differentially accepted.

> The caste distinctions are actually gulfs which divide the populations into antagonistic camps... The caste line—or, as it is more popularly known, the color line—is not only an expression of caste differences

and caste conflict, but it has come itself to be a catalyst to widen differences and engender conflict.[5]

For those exploring the nature and dynamics of social differentiation there are a number of common themes that have become especially important in community analysis. One concern is over the nature of the strata and the vertical differentiations that distinguish one class, status, caste or group from another that form a hierarchy. Some of the questions being asked are: (1) how many strata are there? (2) what is the basis or criteria of a person being placed in one strata or another? (3) how rigid or flexible are the lines? (4) do the strata change and what factors seem to bring about these changes?

A second concern is over the relationships between the various strata. Some of the questions being asked are: (1) what is the relative status, rank, or value placed on each stratum? (2) how does each stratum perceive and behave toward others? (3) can an individual move from one stratum to another and how difficult is it to do so? (4) how accessible are the governors? A third concern involves the search for integrative mechanisms. Here the questions asked are: (1) what vertical lines exist that may provide opportunities for persons within a group to overcome differences posed by the horizontal strata lines? (2) what social institutions and associations facilitate cutting across strata lines?

A final set of concerns involve the socioeconomic and political consequences that flow from various patterns of social differentiations. Social analysts have attempted to discover what difference it makes to individuals and groups to be members of one class, status, caste, or group. Some analysts have focused on land use and spacial patterns, others on residential neighborhoods, and still others on quality of life indicators.

Most contemporary social scientists have devised measurements that take into account several factors important in gauging social status. Considerable refinements have been made to the basic concepts and their empirical measurement. Duncan, for example, has identified some mechanisms of stratifications as: (1) ascription (status conferred by birth), (2) inheritance, (3) genetics, (4) socialization, (5) access to opportunities, (6) environment, and (7) differential associations.[6] He lists some of the bases of status as: (1) wealth, assets, and property; (2) level of living and possessions,

(3) prestige, honor, reputation, fame, and esteem; (4) education, knowledge, and skill; (5) style of life, status symbols, manners, and language; (6) power and authority; (7) legal status and freedom; and (8) welfare.[7] (See the comparable list of resources used in community decision-making in Chapter 5.)

The most common approach is to use the concept of *socioeconomic status*, commonly referred to as SES.[8] This is a measure of three factors: occupation, income, and education. By classifying the residents of a town according to these factors, one can identify the kind and degree of social stratification in the community. There are those who say that income is so closely related to the ability to obtain a good education and qualify for a good job that it is not necessary to look any further than income as a measurement of social status. Yet that is not always the case, as improvements in one's job and/or education do not necessarily mean an increase in income.[9] For example, the highly educated teacher does not earn as much as, or little more than, the less-educated craftsman. The use of SES should help discover the gulf between the top and bottom of the local social structure, and the point or points at which most residents are clustered.

Small Town Differentiations

These ideas and others on social differentiations have taken hold in the United States and stimulated some people to question the reality of the American belief in equality, particularly as it appears on the local or community level. One may wonder: why bother about a town's social structure? There are two reasons, neither of which will be unanimously endorsed by all observers, nor will they be totally rejected. First, social structure may be the key not only to community analysis but also to a town's prospects and future course of change and development. Issues around community improvement may be more conflictful in a highly stratified small town, with sharp divisions between upper and lower strata, than in a less stratified community. A highly differentiated town might engender more discussion, less consensus, and a greater variety of ideas for community betterment than a town with little social diversity. Some social scientists go so far as to suggest that a community's social structure is the primary generator of local cultural norms and local patterns of influence. In fact, they conclude that if you know the class relations you know nearly all you need to know about the character and dynamics of community change.

The second important feature of a small town's social structure is related to the size of the community. In a small town, unlike a large city, almost all of the residents recognize each other by sight, and in many cases they know each other more intimately. People experiencing the personalized interactions of small town living believe that they have a fairly accurate sense of the social status of the other residents. Often there is a high degree of agreement and residents will discover that their perceptions as to who is upper or middle class are shared with others. For example, the banker in town will be generally considered upper class, and the "reliefers," lower. However, in a big city social rankings are not as clear because occupations are less obvious. Perceptions of social status frequently become the basis for interpersonal relations among small towners. This apparent broad agreement on social status sometimes breaks down when it is questioned. A few pointed questions might reveal that one resident ranks people by income level, another by job position, and a third by life-style. Occasionally the perspectives of different strata do not agree. The upper stratum, for example, might perceive small differences among its peers, but then lump all other residents into one or two broad categories. Similarly, lower-status groups might simply classify everyone above them as "rich." These distinctions in perspective, along with the beliefs, behavior, and actions that result from them, are components of a town's social structure.

During the 1930s and 1940s, Warner, an anthropologist, was one of many who began looking at community class structures. In his study of two communities, "Jonesville" (population 6,000) and "Yankee City" (population 17,000), Warner discovered that residents were aware that they ranked their neighbors according to criteria more varied than those used by Marx and Weber. Warner states: "while occupation and wealth could and did contribute greatly to the rank-status of an individual, they were but two of many factors which decided a man's ranking in the whole community."[10] When he asked the residents to describe and rank people in the community according to the way they perceived them, he discovered six levels of classes (percentages from "Yankee City").

(1) Upper-upper: (1.4 percent) the local social "aristocrats," members of families with long histories of wealth and social standing, who tended to stay within their own circles
(2) Lower-upper: (1.6 percent) another wealthy group, in some individual cases, even wealthier than members of the upper-upper class, but lacking the properly distinguished backgrounds necessary for belonging to the higher group
(3) Upper-middle: (10.2 percent) mostly businessmen and professionals
(4) Lower-middle: (28.1 percent) "white collar" clerks and small businessmen
(5) Upper-lower: (32.6 percent) workingmen and small tradesmen
(6) Lower-lower: (25.2 percent) those associated with lower incomes and poverty life-styles, usually residing in the less desirable parts of town

Davis observed a wide variety of social perspectives that members of each stratum held toward the members of their own and other strata. For example, Figure 3.1 illustrates how people in each of the strata perceived themselves.[11]

All strata except the lowest held negative views of the lowest strata by referring to them as "po' whites," "no 'count lot," or "shiftless people." The upper class tended to idealize the past and lineage, the middle class emphasized the importance of wealth and morality, self- and community-improvement, while the lower class lacked integration into the community and emphasized economic insecurity, jobs, and the importance of residential areas.

Vidich and Bensman not only identified five social status sets in Springdale (population 2,500), but discovered each had important and different ideological orientations toward investment, savings, consumption, and work. Table 3.1 illustrates the size and orientation of each strata. They believe "social class" to be replacing "economic class" distinctions and that "the underlying secular

Figure 3.1: Several Self-Perspectives by Strata (old city)

Strata	Self-Perceptions
Upper-upper	"old aristocracy"
Lower-upper	"aristocracy," but not "old"
Upper-middle	"people who should be upper class"
Lower-middle	"we poor folk"
Upper-lower	"poor but honest folk"
Lower-lower	"people just as good as anybody"

Table 3.1: Social Class Ideological Orientations

Social Class	Percent Population	Investment	Orientations Savings	Consumption	Work
I. *Middle Class*					
1. Independent entrepreneurs	13	−	+	−	+
2. Prosperous farmers	25	+	−	−	+
3. Professional/ skilled workers	9	−	−	+	−
II. *Marginal Middle Class*					
1. Aspiring investors	10	+	−	−	+
2. Economic/social immobile ritualists	10	−	−	+	−
3. Psychological idiosyncracies	2	−	−	+	−
III. *Traditional farmers*	10	−	+	−	+
IV. *"Old Aristocrats"*	1	−	−	−	−
V. *Shack People*	10	−	−	+	−

Key: + adopted orientation
− orientation not adopted

SOURCE: Adapted from Vidich and Bensman, *Small Town in Mass Society*, pp. 49-78.

trend indicates there is a shift from production to consumption values in the community."[12]

In a study of caste and class in Old City (population 10,000—half black and half white), Davis and his colleagues discuss the social matrix:[13]

> Life in the communities of Deep South follows an ordered pattern. The inhabitants live in a social world clearly divided into two ranks, the white caste and the Negro caste. These color-castes share disproportionately in the privileges and obligations of labor, school, and government, and participate in separate families, associations, cliques, and churches. Only in the economic sphere do the caste sanctions relax, and then but for a few persons and in limited relationships. Within the castes are social classes, not so rigidly defined as the castes, but serving to organize individuals and groups upon the basis of "higher" and "lower"

status, and thus to restrict intimate social access. Both the caste system and the class system are changing through time; both are responsive to shifts in the economy, in the social dogmas, and in other areas of the social organization. Both are persisting, observable systems, however, recognized by the people who live in the communities; they form Deep South's mold of existence.

America is a pluralistic nation made up of many people with many different ethnic identities based on nationality, religion, and race. While the ports of entry were the seaports and big cities, various ethnics have settled throughout America in small towns and neighborhoods. Each ethnic group may be criss-crossed by social class, and contain its own primary groups of families, cliques, and associations. Gordon has argued that the United States

> is a multiple melting pot in which acculturation for all groups beyond the first generation of immigrants, without eliminating all value conflict, has been massive and decisive, but in which structural separation on the basis of race and religion ... emerges as the dominant sociological condition.[14]

In fact, Gordon has formulated the term "ethclass" as the intersection of the vertical stratifications of ethnicity with the horizontal stratifications of social class.[15]

In recent years social analysts have examined ethnic relations as a supplement to social class to discover various patterns of intergroup relations. These include the study of prejudice and discrimination and their impact as well as various efforts to form cross-ethnic coalitions and to cope with intergroup stress and tensions. Williams has identified the main variables that are related to intergroup contact and interaction that lead to cooperation or conflict[16]:

(1) *status attributes* of individual persons—age, sex, education, prestige, rank;
(2) *culutral values* held by individuals;
(3) *stereotypes and prejudices;*
(4) *Personality* structure and dynamic;
(5) characteristics of *situation of contact* relative to collective goals, role of third parties, number of participants, and proportion of each category;

(6) *collective properties emerging from interaction,* i.e., awareness of shared attitudes, sense of group position, patterns of concerted action, and group cohesion.

Reissman has compiled the results of many small town community studies into one hypothetical community, "Hometown."[17] He believes that classes as they are defined by the objective criteria of income and wealth on a national level are real and important in small towns. Friendships and social activities seem to occur within narrow class boundaries, with little interaction occurring among class groups beyond that necessary for day-to-day survival. For some people in town, there is a sense that it is a matter of "my class (or people like me) and all those other people." The larger the gulf among classes, the less the members of each class appreciate the diversity within the other classes. One symbol of class is found in the choice of residential location, such as living "up on the hill" or high-status part of town in contrast to the lower-class area referred to as "the other side of the tracks."

Just as neighborhoods are relatively clearly defined, so is the different behavior of each class. The middle class, frequently the largest of all, tends to be characterized as hard working, clean living, and very involved in civic affairs-type of people. The lower class is stereotyped, though frequently unjustified by the facts, as being lazy, shiftless, and immoral. Little recognition is given to the fact that many poor people in small towns strenuously try to emulate the life-styles and values of the middle class. The upper class tends to remain socially distant and aloof from the daily affairs of the town, its "snobbery" inhibits its involvement with the other classes. Local politics is usually left to the civic-minded middle class, except when an issue arises that might strongly affect the interests of the upper class. Members of the upper class are known to dominate decision-making. Reissman believes, based on the conclusions of a number of studies, that the first response of small towners to the idea of stratification—"we don't have classes here," or "everyone's just plain folks"—is just a veil hiding an active and generally agreed-upon class structure.

Spacial Aspects of Social Structure

Small towners do not go about systematically measuring the social stratification in their everyday lives by using the SES concepts of social scientists. One concrete expression of stratification that small towners quickly perceive is the differences between residential neighborhoods. Although it is possible to hide many other manifestations of wealth and status, one's social position is far less "anonymous"[18] in the choice of neighborhood. People choose residential locations based on preferences that reflect their positions in the social structure. It becomes apparent that land uses are not random, but conform to definable patterns. The clustering of social groups, the characteristics of particular neighborhoods, the location of one neighborhood or land use to another—these are the physical expressions of social stratification in a community.

Social scientists have for many decades examined the similarities and differences within communities according to land use patterns. Studies conducted at the University of Chicago in the 1920s developed an approach known as "human ecology." Though developed initially to explain patterns of human settlement in the city of Chicago, some of the insights the human ecologists developed there are demonstrably important in small communities. One should be careful not to assume that what helped to describe and explain Chicago is necessarily applicable to rural villages and towns.

Human ecologists believe that the growth and development of human communities was comparable to those of plants and animals. That is, like plants and animals, humans tended to settle and use various areas through continuous natural processes. Just as different species of plants are distributed in different areas across a landscape, people and their activities are located at different points throughout a community.

In Chicago, Park and his colleagues found that the shape of the community took the form of concentric rings, each ring comprising a generally common function and character, differing from the content and function of the other rings.[19] Due to the ecological parallel, the rings are referred to as "natural areas." The concentric pattern, as they described it, is characterized by a core or center for business and commercial uses with circles of residential areas moving out in waves from the center. Other communities seem to have other kinds of developments. Hoyt, for example, proposed a different process of sorting out land uses which produced an axial or sectoral pattern.[20] He suggests that rather than forming rings, activities and land uses divide themselves into wedge-shaped quadrants, usually located along radial highway routes that slice the town. One wedge-shaped quadrant or sector might be a residential area, while another might be a commercial one, and so forth. Both the sectoral and concentric patterns focus on one central business district. Later Harris and Ullman discovered that communities are not always limited to one business core district.[21] In some communities, a number of highly specialized and intensive centers of land use can be found, rather than one centrally located core. As opposed to the concentric or sectoral patterns, the "multiple nuclei" or "polynucleated" schemes present a picture of a more fragmented, less uniform community land-use pattern.

Many smaller communities do not initially display a differentiation of uses or a diversity of residential areas that are characterized by one of these complex schemes; however, as they grow in size they tend to develop one or the other of these patterns. In small towns that remain small, land-use patterns are merely divided between one side of the community and the other; sometimes a railroad track, sometimes a stream serves as the dividing line separating residential areas. This division occasionally highlights social differentiation as to who resides where in town. Some sections are known as the older neighborhoods, populated by long-time residents with strong linkages to the town's heritage and tradition. On occasion these residents become the town's "aristocracy," more respected and influential in local affairs than relative "newcomers." Typically among the more stable and long-lasting members of a community are its homeowners. Renters comprise a more transient group, moving with greater regularly and frequency. The distinction between homeowners and renters is also a manifestation of social stratification, and it can be seen that homeowner neighborhoods are usually better places to live. The status differential is not merely a question of transiency; homeowners are also economically and culturally given a higher status position, though in some cases renters may actually be of higher SES.

The ecological approach has its critics. Bernard, for example, notes that most natural areas are not all that internally homogeneous.[22] The closer one looks, the more apparent become differences and variations in just about any area identified as natural by the ecological method. In a totally residential neighborhood, closer inspection reveals sharply different life-styles, incomes, and values, despite the common location and housing type. An area identified for its unique character, such as its uniformly ethnic population or particular social purpose, will yield a number of internal variations. To ensure some sort of common functional definition, either the nature of the land use must be defined very generally so that many variations can be embraced in the same category, or the area or unit of analysis must be so small as to ensure that all residents of the location will conform to the particular characteristic that makes it an ecologically natural area.

In many communities, the geographic expression of social stratification is preserved through municipal land-use plans and zoning regulations. The objective of such plans is to control the development of a community so that undesirable or incompatible uses of land are not located next to residential neighborhoods. The kind of use and pace of growth will be regulated in order to minimize disruption of the town, and the sensitive environmental features of the area will be protected from irreparable damage. In addition, in many communities, zoning serves as a measure to enforce the homogeneity of neighborhoods—for example, a neighborhood of single-family homeowners will be zoned to prevent the construction of multifamily rental housing, low-income housing, and the like, which will preserve the social character of the neighborhood. In that way, zoning becomes an exclusionary tool to prevent some people from freely choosing the neighborhood for their home. It also prevents the mixing of low-income and minority residents with others of higher status.

One common classification of land use is the categorization of land into "public" and "private." Most planners' approaches to classification are to identify which properties are owned by the municipal, state, or federal government and which are in private hands. For the purpose of social stratification, the public/private distinction, if based solely on ownership, is inadequate. Many private properties are perceived and used as though they were public. On the other hand, in some towns, designated public areas are actually not used by all social groups as some segments of the community feel unwanted and excluded from some territories in many cases, as intended by other, dominating social groups. In examining the ecological nature of the public/private dimension of land use, the following points are relevant: (1) who owns the land? (2) who uses the land? (3) who decides who uses the land? Answers to these questions reveal that "public" territories, such as parks, playgrounds, ballfields, community centers, and public buildings are not always open and accessible to all, but perceived and used differently by various social groupings in the town.

Lynch has interviewed residents to record the various images of parts of the community to determine the land-use patterns. He stresses elements of form—what elements of the community's use of land stand out as important features in the daily activities in the community. Residents use these features as guideposts for orienting themselves in the community, turning the town from a maze of streets and houses into a pattern of meaningful components. As a professional planner, he organizes the community into five categories:

(1) paths: "the channels along which the observer customarily, occasionally, or potentially moves"[23]; ranging from major thoroughfares to backyard alleys, the sense of movement is a primary tool of many people for orienting themselves in a community;

(2) edges: "boundaries... linear breaks in continuity ... barriers, more or less penetrable, which close off one region from another ... or ... seams, lines along which two regions are related and joined together"[24];

(3) districts: areas of the community "which have some common character"[25] perceivable both from the inside and outside (as a resident of the "district" from its surrounding environs);

(4) Nodes: cores or concentrations of activity and purpose, ranging from major areas of converging "paths" to important focuses of activity and identity in "districts"[26] (such as a location commonly used for several gatherings, such as a popular store, bar, street corner, enclosed park, and the like);

(5) Landmarks: "a rather simply defined physical object" that is "unique" and "memorable,"[27] "easily identifiable" or functionally prominent, that is significant to community residents.

Whole towns or just neighborhoods can be viewed through this framework.

Lynch found social class to be important as residents classify areas of town. One might also expect different social groups to have differing perceptions of a community's paths, nodes, districts, and edges. Though on the surface these categories appear physical, what distinguishes a path from any other road, or a landmark from any other building, or an edge from any other boundary, is its social meaning and social use. Different social groups are likely, for example, to identify edges in very different manners—in many cases, according to where they feel comfortable and accepted and where they feel excluded or resented. Lynch's own studies of American cities are themselves somewhat biased in favor of the perspectives of the middle classes. Care must be taken to explicitly recognize and understand the social dimensions of the elements of community form.

Socioeconomic and Political Consequences of Social Differentiation

In almost all communities the social structure plays an important role in governing the way residents behave and interact. The significance of learning that in one small town the poor live in "Riverbrook" and the rich on "Hill Street"[28] is not just a matter of geography. Such factors as: (1) social distance, mobility and the deference to prevailing community values and customs; (2) wealth and income which provide the basis to participate in the marketplace and enhance one's quality of life; and (3) the mobilization of influence to affect the collective decision-making process that determines the availability and delivery of public services are involved in the distinction made.

Social Interaction

Living in a small town almost all members have occasion to interact with one another—certainly more so than those who live in a big city. The chances of meeting one another has a high probability as one shops, works, and plays in virtually the same institutions. There are even occasions when people will choose to join members of other social groups for various noncompulsory activities—the school teacher and the bricklayer might sit down for a friendly game of cards or a drink, especially if they live in the same neighborhood. The occasions for two people of different status backgrounds to socialize together are, however, less than they are for two people of similar status positions. The greater the social gap between persons, the less likely they are to interact. The landlord will collect his rent from the tenants of the building, the day laborer will take his orders from the shop foreman or plant manager, the reliefer will fill out the government welfare forms handed to him by the welfare caseworker, but after the formalities are said and done, the likelihood of socializing and friendship between pairs goes down, despite the exceptional cases whose rarity tend to prove the rule.

Members of particular groups in general tend to keep to their own. They seem to avoid living too near different social classes, they intermarry less frequently than they marry their own kind, and they seldom go out of their way to socialize with members of other strata.

More Americans experience greater mobility than members of any other country. They not only move from place to place, but move vertically, involving status changes. They do so between one generation and the next as well as within one generation or "career" advancement. The main determinants of mobility are: (1) technological change; (2) differential rates of reproduction or migration; (3) changes in abilities; and (4) changes in attitudes. Mobility, of course, is "immensely variable."[29] While some move upward, others move downward, voluntarily and/or involuntarily. Warner and Srole not only found various ethnic groups based on nationality, religion, and race to have different rank-status, degrees of subordination to the dominant Yankees, strength of a group subsystem, but varying time to assimilate and acculturate into the mainstream of community life.[30] Blacks were very slowly assimilated into a color caste status based on race. The processes for mobility were the "acquisition of material symbols, including residences, increased occupational status, extension of formal and informal relations in the society, and change in behavior modes."[31]

For a member of the black caste to cross the "color line" is difficult and is referred to as "passing." Myrdal notes that "passing requires anonymity and is, therefore, restricted to the larger cities where not everyone knows everyone else. A Negro from a small community can pass only if he leaves the community."[32] Bernard gives three cogent reasons why the black elite reject the common methods of upward social mobility as dysfunctional for black people: "1) the costs of individual upward mobility in terms of mental and emotional health were excessive; 2) the costs to the black community in terms of class schism were excessive; and 3) the vulnerability to exploitation by the white world was enhanced."[33]

Quality of Life

The ability of a family to enter the marketplace and acquire the necessities of life and whatever other amenities it prefers is largely dependent on the family's wealth and income. This is especially true when it comes to the purchase of shelter, food, clothing, and medical care, all in the private sector. However, a garden in the backyard might supplement one's food bill. Government payments for those in poverty and the aged help assist some to receive goods and services they could not otherwise attain.

Nonetheless, earned income remains a central measure of socioeconomic status. As such, one should expect it to be a fairly good predictor of a family's purchasing power. Table 3.2 indicates that the lowest quintile (20 percent) of families earned 5.4 of the nation's income in 1970, while the highest quintile earned 40.9 percent. This means that the bottom group had only 5 percent of the nation's earned income, while the top group had 8 times as much. Surprisingly, this pattern has remained fairly constant throughout the post-World War II period.

The tax bite on earned income is shown in Table 3.3. It indicates not only a constant trend during the post-World War II period, but shows surprisingly little difference between each group

Table 3.2: Percentages of Aggregate Income Received by Each Fifth of the Nation's Families

	1950	1960	1970
Lowest fifth	4.5	4.8	5.4
Second fifth	11.9	12.2	12.2
Middle fifth	17.6	17.8	17.6
Fourth fifth	23.6	24.0	23.8
Highest fifth	42.7	41.3	40.9

SOURCE: Adapted from U.S. Statistical Abstract, 1974, Table No. 619, p. 384.

Table 3.3: Percentage of Personal Income Taken by Federal, State, and Local Taxes from Each Fifth of Consumer Units

	1948	1954	1965
Lowest fifth	26	24	22
Second fifth	25	26	24
Middle fifth	26	27	25
Fourth fifth	26	28	26
Highest fifth	32	36	33

SOURCE: Adapted from Council of Economic Advisors (mimeo), 1969 *Annual Report*, p. 161.

except for the highest fifth. One must remember that the family's tax responsibilities not only include such progressive taxes as the federal income tax, but such regressive taxes as the state sales tax and the local government property tax.

The income gap between blacks and whites is further compounded. Table 3.4 contrasts some economic statistics for those living in nonmetropolitan areas. The statistics not only show black families earn about half that of whites but also the proportion of blacks in poverty is three times that of whites. The discrepancy of income prevails for those blacks holding the same occupational status as whites. This not only means less money to buy goods and services but reveals the persistence of the caste-color line, despite the efforts of the federal government in the 1960s to redress the basic conclusion of the Kerner report on racial disorder: "our nation is moving toward two societies, one black, one white—separate and unequal."[34]

With less income it is obvious that those in the lower-income strata can buy less. Miller and Roby have pinpointed some contrasts between the poor and the nonpoor[35]:

(1) 62 percent have no savings compared to 60 percent of those earning $15,000 or more who had savings of more than $6000;
(2) only 50 percent have life insurance compared to 98 percent of others;
(3) two and a half times more live in substandard housing;
(4) 46 percent own their homes compared to 86 percent for the highest income;

Table 3.4: Comparative Financial Statistics for Those Living in Nonmetropolitan America – 1970

Item	Total	Black
Median family income	$7832	$4060
% poverty families	15.4	46.6
% receiving public assistance	18.7	19.8
% 65 years & older receiving Soc. Security	75.0	61.8
% poverty families with house lacking some or all plumbing	25.9	55.6
Males		
Median income	$6403	$3458
Professionals	9130	6199
Craftsmen	7124	4148
Operatives	5963	3920
Laborers	3972	3169
Farmers	4731	1579
Farm laborers	2462	1807
Females		
Median income	$3107	$1784
Clerical	3573	2573
Operatives	3405	2798

SOURCE: *United States Summary, 1970, Census of Population, General Social and Economic Characteristics.*

(5) spend more of their income on housing, have less home insurance, and are less able to maintain or enhance their housing investment;
(6) receive two and a half times *less* public subsidies;
(7) make one-third less dental visits but require four times more teeth extractions;
(8) had lower rates of hospitalizations but were required to spend more days there;
(9) experienced the highest rates of diseases—heart, diabetes, arthritis;
(10) had higher infant mortality rates.

Mobilization of Bias

Virtually every study has shown a high correlation between socioeconomic status and political participation. In small towns, Hamilton found not only much lower rates of participation by working class (53 percent voting compared to 78 percent for the upper-middle class), but also less interest in campaigns (19 percent compared to 61 percent for the upper middle class). He found that social class differences are greatest in small towns with workers and lower-middle class members very much "out of it" and the upper-middle class members very much "in." He suggests,

> that in the small towns, politics, so to speak, "takes care of itself." Things proceed in their traditional way with little or no special efforts being necessary to "mobilize the masses." Since there is virtually no challenge from below, there is little need for the community leaders to undertake any special organizational efforts.[36]

Higher rates of political participation in conjunction with greater participation in the community's voluntary organizations by the upper strata tend to facilitate their mobilization of bias. For example, Warner found in a summary of small town educational systems that 94 percent of the school board members were from the upper-middle and upper strata, while 94 percent of the teachers were from the upper- and lower-middle strata. The student body, on the other hand, represented the following socioeconomic statuses: 10 percent from upper and upper middle, 30 percent from lower middle, 60 percent from lower.

Thomas, citing the typical school's bias in favor of upper and middle over lower strata students, explains that the control of the school's policies and daily operation by this upper strata boards

> favor patterns of behavior subscribed to by their own social strata and to disapprove of many characteristics of lower-class pupils. Furthermore, parents of children in the higher strata are apt to exert greater influence over school personnel than are lower-class parents.[37]

They do so, Thomas asserts, by: (1) provision of better facilities; (2) preferential classroom treatment; (3) curriculum materials

Table 3.5: The Socioeconomic Status of Leaders and Rank and File

Status	Leaders (N = 81) %	Rank and File (N = 1104) %
Upper	84	14
Middle	16	62
Lower	0	24

SOURCE: Robert Presthus, *Men at the Top* (Oxford University Press, 1964), pp. 181, 184.

suited more to upper and middle classes; and (4) more prestige accorded college preparatory courses.[38]

Presthus specifically demonstrates the vast socioeconomic status difference between the community leadership and the rank and file (see Table 3.5). No leader was a member of the lower class in the two small towns he studied.

Social Structure and Community Change

While most observers contend that small towns are tranquil and consensual, there are times and places when social change does occur. In a constantly changing larger world it is difficult to maintain the status quo, especially if external forces—social, economic, political—are penetrating the local autonomy. The sharp class distinctions that are normally associated with social distance tend to wane when communities become influenced by the wealth, values, and institutions of the larger society.[39] Internally, however, change in a specific community can be in large measure stimulated, regulated, or resisted by those in power, and the extent or form of that change depends on their ideologies or value preferences. Change can be progressive or regressive; it can be incremental or rapid, and it can be orderly or conflictive. To some extent, community change stems from the value differences of the various strata in town and how the members of various cleavages react to each other in formulating a community's policy directions. Here we will speculate on the dynamic interactions which may modify the social structure of the community.

Schnore distinguishes those aspects of social status that can or cannot be changed and whether the change is reversible or irreversible.[40] For example, changeable items such as age and education are irreversible, while citizenship, occupation, income, religion, place of residence, and marital status are reversible. On the other hand, one's place of birth, family connection, and race are unchanging and irreversible.

As between the various strata, there are several possible modes of interaction. One is simply *coexistence*—each status group makes no effort to interact with another and status differentials are played down. A second is *neglect* or *avoidance*—one status group deliberately refuses to acknowledge the existence of another group, a common practice in a caste-based community. A third is *cooperative* interaction where the various strata join together to enhance the common good of all, believed to be the prevailing mode in most small towns. A fourth is *competitive* where each status group attempts to achieve its own goals and objectives generally at the expense of other status groups. The final mode is *conflictive* as each group perceives the other as threatening and antagonistic. Gamson distinguishes between "conventional" and "rancorous" conflicts.[41] In the former, opponents regard each other as mistakenly pursuing a different but legitimate goal, while in the latter there is the belief that community norms are being violated in waging political conflict.

The circumstances that generate these modes of interaction between or among strata are largely situation-specific; that is, where conflict may be the result of conditions in Small Town A, cooperation might occur in Small Town B because of a totally different set of factors. For small towners, recognizing the tone of intergroup interactions and relating it to the social conditions of the community may be a critical consideration in planning for community development.

The separation of cooperation from competition among social strata is difficult. It is possible to cooperate while being competitive. In fact, the American system rests on just such a notion. For example, two competing social groups will agree to work together based on the mutual expectation of rewards.[42] In other words, though the groups have different objectives and are basically competitive, they will accept a limited amount of cooperation in the hope that each group's goals are simultaneously pursued. The Hill Streeters and the Riverbrookers may both agree to an urban renewal project, the Hill Streeters with the expectation that the town will be spruced up by the elimination of slum housing, and the Riverbrookers with the assumption that they will be relocated at the government's expense to more desirable housing. With mutual expectations of reward, even the most antagonistic of groups can find a means of temporary accommodation, each achieving its particular goals within a community context.

Competition does not always lead to cooperative working arrangements, however. Sometimes the interclass relations degenerate into bitterness and strife. A common manifestation is to blame the bad conditions of the town on the other social strata, a form of "scapegoating."[43] When it comes to scapegoating, accuracy is not as important as convenience—it is the easy way out to blame the town's problems on some other social class. Often, the scapegoat is the poverty-stricken "reliefers" in the town, or sometimes a particular ethnic or racial group. Many a small towner has witnessed his neighbors and friends persecute others for less than justifiable reasons, in the end driving minority social groups from their rightful place in the community. Frequently, however, the elimination of the scapegoat leaves the town with about the same problems as before except that now the residents must find a new scapegoat. According to Beshers, social distance is the key to scapegoating. Were it not for social distance, people would have little basis for pejorative and biased conceptions of social classes other than their own.

It is not too difficult to see how the tensions between members of different social strata can eventually become conflictful. The language of scapegoating can, with the right combination of factors, quickly degenerate into conflict. Lower status people may view the upper classes as "exploiters," and may attribute the lack of social progress in their lives to the domination of those in better straits. From the point of view of the upper classes, those below them represent threats to their bequeathed or, in some cases, hard-won privileges and rewards. Often people of a particular social group may view their nearest competitors as "enemies," although the more meaningful source of their competition for scarce resources is actually several status sets higher.

Most conflict tends to sharpen "class" identities. However, much of American life is shaped by a trait peculiar to the development of this nation. Whether laborer, clerk, professional, or small businessman, the bulk of Americans consider themselves "middle class." Much of that sentiment is influenced by the mass media—the images of small towners seen on the television portray role models that are commonly defined as "middle class," and the tendency is to identify with those middle-class role models. If that is so, relative differences in SES may not be as divisive a factor as other issues: race, ethnic group, age, life-style, and so on.

There is a dimension of territoriality to the way social groups settle in a community—that is expressed as "natural ones" in the ecological approach, or "districts" in Lynch's scheme. The dynamic aspect of the stratification of neighborhoods is in the way neighborhoods change. A neighborhood composed of one social class may suddenly find itself undergoing a change as members of other, competing social groups begin to move in. In a sense, the neighborhood is *invaded* by another social group. Frequently, the diversity introduced by new residents is intolerable for existing residents, and they respond by leaving and reestablishing their homogeneous district at another site. The process where the original residents move out, and a new social group takes their place, is known as *succession,* the result of a successful *invasion.* Where the invasion of a new group does not succeed, the existing natural area has resisted the challenge and preserved itself as *segregated,* according to whatever criterion distinguished that social group from others.

The terms used to describe the dynamics of human ecology in land use do not have pleasant connotations. Invasion, succession, and segregation all ring of conflict. Within a social system, social differences have the potential for producing tensions and intergroup hostilities. The result is not necessarily (Marxist) class warfare. But social differences can and do produce social conflict. Certain patterns of stratification are potentially more conflictive than others. Clark has developed a rough typology of social stratification systems, based on the various shapes possible in a social hierarchy. With some modifications in shapes and terminology, the forms of stratification described by Clark are applicable to small town social systems.[44]

Where one or only a very few individuals dominate the social structure, a pyramid is the most likely shape of the social structure (see Figure 3.2). As in a one-company town, this type may be characterized by the relative harmony of paternalistic domination by the company owner. The potential for conflict is very high if the social imbalance is felt to be particularly oppressive. The trapezoid is like the pyramid, except that its upper point does not exist (see Figure 3.3). Unlike the pyramid, with an individual or one-family at the top, here many more individuals may be members of the upper echelons. Also, the difference between the upper and lower strata does not loom quite so large. It is obvious that this system might experience less rancorous conflict than others.

The rectangular stratification system may be the most stable of all the forms (see Figure 3.4). It is characterized by a large middle class, with a relatively small gulf between the top and bottom strata. The hourglass has a large number of persons at the top and

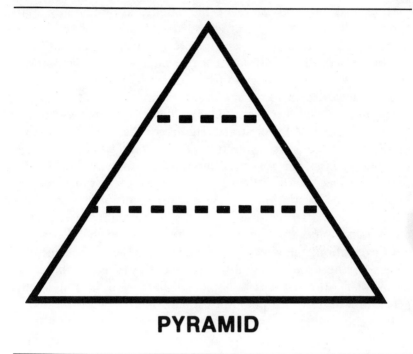

Figure 3.2

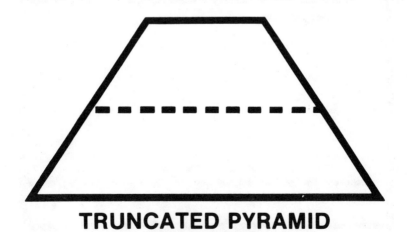

Figure 3.3

bottom and relatively few in the middle as potentially quite conflict-ridden (Figure 3.5). The sharp contrast between those who are well off and those who are not has the potential for interstatus conflict. Without a large middle class, community antagonisms are more likely to proceed unresolved.

Two factors seem to be important in estimating the stability of community social structures. One is the relative proportions of upper, middle, and lower strata. The larger the middle group, the more stable the social structure. Another factor of seemingly paramount importance is social mobility. If residents feel stuck in their positions, with little or no chances of advancement or improvement, frustration and the potential for conflict can brew. Where there is a strong possibility of social mobility or improving one's social status through gaining a better job, higher income, or more education, the conditions for social conflict are lessened. Of course, if the upward mobility of some groups is interpreted to mean that other groups will plunge downward in status, that perception of downward mobility could stimulate antagonisms.

It is quite difficult to generalize about the social structure of small towns in America. For most of what we know has been based on a case by case study of only a few of the thousands of

Figure 3.4

Small Town Social Structure 107

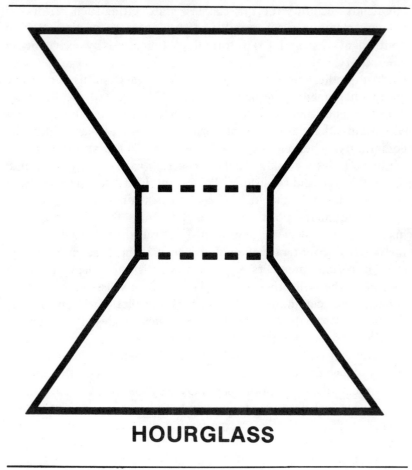

Figure 3.5

small communities. If we turn to the intellectuals and reformers of the progressive period who were born and raised in small towns when America was becoming urban and industrial, we perceive a portrait where

> class distinctions existed, but were typically underplayed by the townspeople. The ethos of most towns was relentlessly egalitarian: the inhabitants liked to think they were too democratic to practice social exclusiveness and too friendly not to mingle freely with those above and below them in the occupational ladder.[45]

Quandt accounts for this classless state, citing such factors as: (1) considerable social mobility; (2) widespread participation in political affairs; and (3) informal participation by everyone in community activities.

Most studies on the American caste, class, status, or ethnic group phenomena have been based on the study of specific communities and then on the smaller ones at that. These studies consistently have found some form of social cleavages within the community whether they be based on: (1) castes—black and white; (2) class—working and managerial; (3) status—upper, middle, and lower; and (4) ethnic—Protestant, Catholic, and Jewish, or the hyphenated Americans such as Mexican-American.

The proximate position of small towners to each other may make the lines of cleavage most distinct but there are a number of factors that counteract their divisiveness. One is social mobility, even if certain members—especially youth—must leave town. A second is the collective representations of myths, symbols, rites, rituals, and ceremonies. A third is conflict with an outside "enemy." A fourth, individual psychological modes of adjustment, was proposed by Vidich and Bensman: (1) the repression of inconvenient facts; (2) the falsification of memory and the substitution of goals; (3) the surrender of illusions; (4) the mutual reinforcement of the public ideology; (5) the avoidance of public statements of disenchantment and the exclusion of the disenchanted; and (6) the externalization of the self through work, sociability, passivity, automatization, and generalized anxiety.[46]

The full applications of these many and assorted adjustive mechanisms would tend to indicate that interstatus relations would be less stressful than those found in diverse big cities. The general sense is that small towns are consensual and change is well modulated. In general, smaller communities are characterized by a high degree of consensus and informal controls which can be expected to develop status systems in which family background, conventional behavior, and social contacts will be important determinants of general status. Blalock offers the following propositions on the relationship between discriminatory behavior and status consciousness[47]

(a) increase as the minority's *average* status decreases;

(b) increase as the minority *individual's* status decreases (intraminority comparison);
(c) increase as the visibility of the contact increases;
(d) be greater for racially distinct minorities than for ethnic minorities, provided there is a high degree of anonymity;
(e) be greater for occupations having conservative norms than for those with liberal norms;
(f) increase with decreasing orientation to the profession, where occupational norms are more liberal than community norms (and the reverse where community norms are more liberal);
(g) increase with the relative importance of social contacts as a determinant of overall status;
(h) be greater for white-collar than for blue-collar occupations;
(i) increase (up to a point) as the general prejudice level in the community increases;
(j) decrease as community size and heterogeneity increase.

Conclusion

Community analysis should provide those about to engage in community renewal some insight into the social structure of a town, the key social groups and the on-going relationship that exists among them. The socioeconomic profile of the residents should reveal the degree to which the community is more or less homogeneous and the major characteristics of the social profile. One should expect considerable consensus to develop regarding renewal in a homogeneous town. Disagreements over the direction and extent of the renewal project is more likely to develop in a community where there is diverse social status groups, especially if the project is perceived as affecting differentially the "interests" of one at the expense of another, particularly at the expense of the dominant group.

Thus, it behooves the advocates of community renewal, whether they be external agents and/or local activists, to make an effort to learn something about the social structure and the more probable impacts of intended community action. Certain kinds of renewal projects will be more acceptable than others, and to some groups more than to others. That is, the physical and economic redevelopment of downtown may be more acceptable to most groups, whereas human renewal of those in poverty or governmental reform may create disagreement because the projects may change the status positions of many and bring about longer-range consequences which are believed to be unacceptable to those who prefer things pretty much as they presently are. Similarly, certain approaches may be more effective than others because there is an accepted way of doing things in town and any change in procedure may result in the loss of influence of one status group to the benefit of another. That is, on-going community decision-making organizations—governmental and voluntary—guard jealously their positions of power when faced with the challenge of neighborhood-based problem-solving groups which present their grievances at a community-wide forum before a television audience.

Most community renewal activities do not have as their intention to change comprehensively the social structure of a town. Instead, there is a tendency to focus on single, more limited aspects of the community. Therefore one is likely to believe that there will not be much change to the local patterns of social status nor to the relationships between groups as a result of various renewal activities. For example, Bruyn found in four small towns that the *power elite* gave some initial support to the program, the *upper status groups* provided steady active support, and the interest of the *working class* lessened with time.[48] He concludes:

> The nature of the class system and the nature of the action program seemed to contain polar interests, one seeking to divide the community, the other to unify it. . . . The philosophy of the action programs emphasizes democratic equality; the class system inherently emphasizes inequality. . . . In two communities there were clear indications that the status of a number of *individuals* in the community was raised through leadership, but the main class structure continues unchanged.[49]

DISCUSSION GUIDE

I. What Are the Indicators of Socioeconomic Status Differences Among Residents of the Town?

From census data contained in the *General Social and Economic Characteristics* volume published by the U.S. Bureau of the Census, information on SES categories can be obtained for towns as small as 2,500 in each state. However, for smaller communities, one should consult with county planning offices, regional planning commissions, and state-wide agencies if data is unavailable locally or in published census volumes. It is possible, then, to classify residents according to the three categories of SES.

Income: Earned income is reported for both families and unrelated individuals. Income levels in a small town can be compared to data for the nation, as presented here.

Income Level	% of All Families in U.S. (1970)	% of All Families in Community
(1) Less than $1,000	1.6%	_____
(2) $ 1,000 - 1,999	3.0%	_____
(3) $ 2,000 - 2,999	4.3%	_____
(4) $ 3,000 - 3,999	5.1%	_____
(5) $ 4,000 - 4,999	5.3%	_____
(6) $ 5,000 - 5,999	5.8%	_____
(7) $ 6,000 - 6,999	6.0%	_____
(8) $ 7,000 - 9,999	19.9%	_____
(9) $10,000 - 14,999	26.8%	_____
(10) $15,000 or more	22.3%	_____

In comparing a community's income distribution with that of the nation, certain important factors may be revealed. One approach is to consider the diversity of resident's incomes. Where do the bulk of residents fall? on the low end? on the high end? Or are they relatively evenly distributed? This defines the town's predominant income groups in terms of numbers. A second step is to determine the extent of the gap between the highest and lowest income groups. If a wide gap exists between top and bottom of the income scale, it is likely that the social structure is highly stratified. It is also important to look for the proportion of residents below the poverty level. In 1970, for example, the poverty line was set at approximately $4,000, below which were 14% of the nation's families.

Occupation: Census materials can also reveal the occupations of those in the labor force, which can be divided between "white collar" and "blue collar" occupations.

Occupational Category	Percentage of Employed Persons U.S.	Community
(A) White Collar		
1. professional, technical, and kindred workers	14%	_____
2. managers, officials, and proprietors	9%	_____
3. clerical and sales workers	24%	_____
(B) Blue Collar		
4. craftsmen, foremen, and kindred workers (skilled workers)	13%	_____
5. operatives and kindred workers (truck drivers, delivery men, and so on)	16%	_____
6. service workers (attendants, housekeepers)	13%	_____
7. laborers (unskilled workers)	4%	_____
8. farm workers	3%	_____

It is important to explore how most of the residents of the town are employed. A blue collar town may be very different from one dominated by white-collar employees. For example, the white-collar community may be

Small Town Social Structure

characterized by higher levels of participation in public affairs than the blue-collar community, but the blue-collar town may be less conflictful.

Education: Census materials report on the number of adults 25 years old and older by the years of schooling they have completed.

	Percentage of Adults Attaining Various Education Levels	
	U. S.	Community
(1) Elementary School		
a. 0 - 4 years	5.3%	_____
b. 5 - 7 years	9.1%	_____
c. 8 years	13.4%	_____
(2) High School		
a. 1 - 3 years	17.1%	_____
b. 4 years (diploma)	34.0%	_____
(3) College		
a. 1 - 3 years	10.2%	_____
b. 4 or more years	11.0%	_____

Educational levels below the fifth grade are tantamount to "functional illiteracy." When the largest proportion of a town's adult population has less than a high school education, the prospects for occupational and income advances are low. Not only is educational achievement an important factor in prospects for the family's socioeconomic advancement, it is quite significant in predicting social mobility (the ability to improve one's status position) for subsequent generations. Studies show that the education of the parents is "more important than the income of the family in affecting how far the youth goes in school."[5 0]

II. What Cultural Factors Lead to Social Differentiation and Stratification?

One should look for other less quantifiable indicators to understand community social structure. For example, in some small towns, the symbols of income, education, and occupation may be displayed openly. It is easy to observe who is better off than others. In other towns these symbols may be less conspicuous. While it may be difficult to tell specifically who is better off, one should look for some persisting pattern. Which of the following factors are an important basis of social differentiations and in what ways:

(1) race_____
(2) ethnicity_____
(3) religion_____
(4) lifestyle_____
(5) age_____
(6) sex_____
(7) neighborhood_____

Many of these factors are not readily found in available census materials. Moreover, the relative "strength" of these factors as impetuses of social differentiation may not be readily apparent merely by tallying figures. One town may have significant income difference, but that may not be as important an impetus of social differentiation as religion, for example. To evaluate the strengths of these indicators, there is no effective substitute for getting out and observing and talking to people. One might ask small groups of people prepared questions designed to draw out residents' impressions of social differentiation. In many cases, the realities of everyday life are better keys to social stratification than census data. A well thought-out interview format might well discover some of these factors, and what is learned may be surprising. To confirm the results of the interview, observation of the behavior of residents may be helpful. The character of membership in social or fraternal clubs, the composition of residents for or against different sides of a public issue, the kind of people who shop at particular stores or frequent certain restaurants or bars—these are also clues to corroborate impressions gained from talking to other residents.

III. What is the Geography of Social Differentiation in the Community?

In addressing this task, the basic criterion to keep in mind is: what areas of the town are used in persistent patterns of behavior, and by whom? It is insufficient merely to identify one corner of town as a "node," and a particular street as an "edge." More important is to define what it is that makes that nodal corner important, what social function the street identified as an edge performs, and so on. Here, for example, are some characteristics that might define the social meanings of the elements of community forms.

Paths: The distinguishing characteristics of paths in the community may be their functions as "exits" or "entrances" or "thoroughfares" or "connectors." In a low-income town, the road leading out of town might be viewed as the "exit" by which those people who are able to improve their social status leave for better places to live and work. In another community, a road may be the "connector" between two distinct social groups, where by traversing that pathway, each knows it is going to meet up with members of the opposing groups.

Some prominent *paths* are:_____

_____

Edges: Edges are often physical barriers, but they also might be social phenomena. The cliché of "the other side of the tracks" suggests a physical barrier, the railroad, that serves as a social divider, placing an edge or boundary on the residential habitats of particular groups. Not all edges are so physically dividing. Some edges are understood to exist even though there may be no sharp physical barrier, such as a line in town beyond which members of one social group do not feel comfortable, or are even made to feel unwelcome. In many cases, communities are given edges by their residents that do not actually conform to the municipal boundaries, such as the case where a wealthy district outside of town is considered part of the community, but a low-income area equally close by is conceptually "cut out" or excluded from the residents' image of the community—a distinction based not on geographic proximity, but solely on social status.

Some prominent *edges* are:_____

_____

Landmarks: Landmarks result from their functional importance (e.g., a firehall), their historic connotations (e.g., an old building serving as a linkage to the past), their physical properties (e.g., an imposing church with distinctive architecture), or their social connotations (e.g., the junk car lot in one end of town or the fancy restaurant elsewhere in the community).

Some specific *landmarks* are:_____

_____

Nodes: These are almost always places defined by their functional significance—places where people tend to gather, where differing social groups interact, or where events generally take place. In a sense, nodes are the locations for significant social "happenings." For example, in one town, the bar across the street from the town hall might be a very important node, where the decisions affecting all social groups in town are made and grievances are brought for consideration by community leaders who informally meet there. That bar would be a more significant node than the village council chambers across the street, where decisions already made in the bar get ratified for formal municipal adoption.

Some prominent *nodes* are:_____
_____
_____

Districts: Districts are intimately related to the concepts of social structure. Though many areas of towns are named for geographic characteristics—"Hilltop," "Riverbrook," "The South End," "The North Side," and so on—the use of these names usually connotes the character of the residents who live in these areas. Residents of Hilltop might be the wealthier social strata of town, while Riverbrook might by synonymous with welfare recipients. The South End might contain a particular ethnic group, while the North End might be known as the part of town where most of the elderly live. Districts are critical instruments for identifying how social groups distribute themselves across a town's landscape.

Some prominent *districts* are:_____
_____
_____

The other important notion of social geography is the question of public or private land use. The first step in investigating this is to identify ownership and designated land use. This may be done by procuring a land use map from the municipal or county planning office. The land use map will generally classify land in town into relatively standard categories: residential, commercial, industrial, public, and undeveloped. If such maps are not available, a visual survey of the town can achieve a relatively accurate version of the same thing, or, to ensure exactness, one could check the books at the town hall for ownership records and compile a land-use categorization in that way.

However, official definitions of land use may not accurately describe their social use. For example, the privately owned plot of land in one part of town may be used in regular affairs as a public territory, for baseball games, picnic lunches, or hiking. Some groups may feel unwelcome in nominally public territories, such as a community park or the town hall. Sometimes, perceived public territories, such as parking lots, are actually privately owned and can be returned to full private status if the groups that use the lots are deemed undesirable. These are the social dimensions of land use that are not reported in ownership records or planners' maps. For this task, like some of the others, interviews, discussions, and observation are the keys to determining what groups use what land in which way.

IV. How Do Members of Different Strata, Status Sets, and Groups in the Community Relate to Each Other?

There are five general modes of interaction available to different social strata:

(1) coexistence: positive noninteraction;
(2) neglect (or avoidance): social groups refuse to deal with or even acknowledge the goals and needs of certain other groups;
(3) cooperation: differing strata working together to achieve individual as well as common goals;
(4) competition: each social stratum tries to achieve its own goals while considering the goals of other groups as threats or obstacles to their own;
(5) conflict: groups find the goals of others so threatening, distasteful, or destructive that they sharply contest other's goals and progress.

Which of the above modes of interaction best characterizes the relationship between various groups in town?_____

_____
_____
_____

On various community issues, these different ways of interacting become clear. Some groups as a matter of course band together and cooperate; others tend to compete for influence and control without ever falling into rancorous hostility; and some habitually find themselves in conflict, unable to work out accommodations or compromises. Finally there are some social groups that seem to be ignored, considered unimportant, and generally left out of community issues. The task here is to define the prevailing mode of interaction among the town's social groups.

NOTES

1. Kaare Svalastoga, *Social Differentiation* (New York: David McKay, 1965), pp. 36-70.
2. Ralf Dahrendorf, "Marx's Theory of Class," in Reinhard Bendix and Seymour Martin Lipset, eds., *Social Mobility in Industrial Society* (Berkeley: University of California Press, 1959), p. 6.
3. Marx termed the first "bourgeoisie" and the second the "proletariat." Other groups, such as the shopowners, peasants, and the like, would eventually be drawn into the "two great whirlpools" of the bourgeoisie and proletariat. (Cf. Dahrendorf, 1959, p. 11.)

4. Leonard Reissman, *Class in American Society* (New York: Free Press, 1954), p. 57.
5. Gunnar Myrdal, *An American Dilemma* (New York: Harper & Row, 1944), p. 677.
6. Otis D. Duncan, "Social Stratification and Mobility," in *Indicators of Social Change*, ed. by Eleanor B. Sheldon and Wilbur S. Moore (New York: Russell Sage, 1966), pp. 683-686.
7. Ibid., p. 687.
8. Charles M. Bonjean, Richard J. Hill, and S. Dale McLemore, *Sociological Measurement* (San Francisco: Chandler, 1967), pp. 379-460.
9. S. M. Miller, Martin Rein, Pamela Roby, and Bertram Gross, "Poverty, Inequality, and Conflict," in Bertram M. Gross, ed., *Social Intelligence for America's Future* (Boston: Allyn and Bacon, 1969), p. 286.
10. W. Lloyd Warner and Paul S. Lunt, *The Social Life of a Modern Community* (New Haven: Yale University Press, 1941), p. 82.
11. Allison Davis, Burleigh B. Gardner and Mary R. Gardner, *Deep South* (Chicago: University of Chicago Press, 1941), p. 65.
12. Arthur J. Vidich and Joseph Bensman, *Small Town in Mass Society* (Princeton, N.J.: Princeton University Press, 1968), p. 78.
13. A. Davis et al., 1941, p. 539.
14. Milton M. Gordon, *Assimilation in American Life* (Oxford: Oxford University Press, 1964), pp. 234-235.
15. Ibid., p. 51.
16. Robin M. Williams, Jr., *Stranger Next Door* (Englewood Cliffs, N.J.: Prentice-Hall, 1964), pp. 360-361.
17. James M. Beshers, "Urban Social Structure," in Murray Stewart, ed., *The City* (New York: Penguin Books, 1972), p. 84.
18. Beshers, *Urban Social Structure*, p. 57.
19. Robert E. Park, E. W. Burgess and R. D. McKenzie, *The City* (Chicago: University of Chicago, 1925).
20. Homer Hoyt, *The Structure and Growth of Residential Neighborhoods in American Cities* (Federal Housing Administration, 1939), p. 14.
21. Chauncey O. Harris and Edward L. Ullman, "The Nature of Cities," *Annals of the American Academy of Political and Social Science* No. 242, 1945, pp. 7-17.
22. Jessie Bernard, *The Sociology of Community* (Glenview, IL: Scott, Foresman, 1973), p. 54.
23. Kevin Lynch, *The Image of the City* (Cambridge: MIT Press, 1960), p. 47.
24. Ibid., p. 47.
25. Ibid., p. 47.
26. Ibid., p. 48.
27. Ibid., p. 48.
28. Warner and Lunt, 1941, p. 84.
29. Anselm L. Strauss, *The Contexts of Social Mobility* (Chicago: Aldine, 1971), p. 253.
30. W. Lloyd Warner and Leo Srole, *The Social System of American Ethnic Groups* (New Haven: Yale University Press, 1945), pp. 283-296.
31. Ibid., p. 77.
32. Myrdal, 1944, p. 684.
33. Bernard, 1973, p. 127.
34. *Report of the National Advisory Commission on Civil Disorders* (New York: Bantam Books, 1968), p. 1.
35. These items were drawn from S. M. Miller and Pamela Roby, *The Future of Inequality* (New York: Basic Books, 1970), pp. 67-118.

36. Richard F. Hamilton, *Class and Politics in the United States* (New York: John Wiley, 1972), p. 262.
37. R. Murray Thomas, *Social Differences in the Classroom* (New York: David McKay, 1965), p. 22.
38. Ibid., p. 23-28.
39. Jessie Bernard, *Social Problems at Midcentury* (Hinsdale, IL: Dryden Press, 1957), pp. 64-65.
40. Leo F. Schnore, *The Urban Scene* (New York: Free Press, 1965), p. 54.
41. William A. Gamson, "Rancorous Conflict in Community Politics," in Terry N. Clark, ed., *Community Structure and Decision-Making* (San Francisco: Chandler, 1968), p. 197.
42. Norton E. Long, "The Local Community as an Ecology of Games," in Roland L. Warren, ed., *Perspectives on the American Community* (Chicago: Rand, McNally, 1973), p. 46.
43. Reissman, 1954, pp. 177-190.
44. Terry N. Clark, "Social Stratification, Differentiation, and Integration," in Terry N. Clark, ed., *Community Structure and Decision-Making: Comparative Analyses* (San Francisco: Chandler, 1968), pp. 27-28.
45. Jean B. Quandt, *From the Small Town to the Great Community* (Rutgers, N.J.: Rutgers University Press, 1970), p. 6.
46. Arthur Vidich and J. Bensman, *Small Town in a Mass Society* (Princeton, N.J.: Princeton University Press, 1968), pp. 292-312.
47. H. M. Blalock, Jr. *Toward a Theory of Minority-Group Relations* (New York: John Wiley, 1967), p. 70.
48. Serveryn T. Bruyn, *Communities in Action* (New York: College & University Press, 1963), pp. 120-121.
49. Ibid., p. 122.
50. Miller et al., 1969, p. 311.

Chapter 4

THE ECONOMY OF THE SMALL TOWN

For all the ups and downs of inflation and recession, the United States is still economically thriving, especially when compared to the rest of the world. An abundance of natural resources and skilled labor makes the North American continent an area of great wealth. However, many small communities have not shared in the nation's prosperity equally with other parts of the country. In fact, many communities are in economic decline. In the words of one scholar, "Rural America is in the process of underdevelopment—retreating relatively and absolutely, in labor force, population, capital inflow, commerce, community structure, and income generation from its pinnacle of a half-century ago."[1]

At one time in the history of the United States, small towns were thriving, prosperous places, able to provide employment, shopping, and homes for their inhabitants. Towns as small as 1,000 people, or even less, offered a variety of goods and services that made the necessity of long and arduous trips to the city relatively infrequent. Now, empty downtowns and deteriorating public facilities in many communities signal hard times for small town economies, as the locus of much economic activity in the twentieth century has shifted to the cities. Once-thriving communities have even been abandoned as "ghost towns" as their economic purposes ceased. It has been one hundred years since Jones wrote about *The Dead Towns of Georgia.*[2] This does not mean, as some have said, that small towns are disappearing from the American landscape. People continue to reside in small towns, and many more desire to do so. But some social scientists have begun to question whether the small town can be a viable economic unit. Simon and Gagnon are pessimistic as they see small towners responding differently to the "crises" of increased urbanization:

> the future is one of improvisation. Their horizons must remain limited— for redevelopment is not only a promise but a threat to the ideologies of small town life. They must lose their best people and business concerns to the larger towns because of greater opportunities, education, and satisfactions there. Those who return will be failures—or be willing, for whatever reason, to settle for what represents second-best in our competitive society.[3]

Adams is also pessimistic that most small towns will not reverse the major trend toward decline. However he identifies two factors that may stabilize or facilitate the growth of small trade centers: 1) centralization or the presence of specialized functions; 2) extralocal orientation and risk-taking on the part of key leaders. He sketches the process of losing local services as follows:

> first professionals, then large dry goods and specialized retail services, then duplicate businesses, and finally the most frequent day-to-day services demanded by the community and hinterland. The further a community goes along the path towards death, the more difficult it appears to be to reverse the process without outside intervention.[4]

Not everyone has written off the small community economically. Schumacher has challenged the unrelenting urbanization of modern society when he reminds us that, "The finest cities in history have been very small by twentieth-century standards."[5] Mishan describes vividly the "overcrowding and uglification" that results from cities growing too large.[6] There is strong evidence around the country that the forces for further "agglomeration" in the cities have begun to abate. Many large firms are decentralizing their operations, opening up branches and affiliates outside of major metropolitan areas. Small nonmetropolitan towns have been the beneficiaries of some of these decentralizing moves. A study of the locational preferences of manufacturing firms in Pennsylvania showed, for example, that between 1961 and 1969 jobs increased at a faster rate in "small center areas" than in "large center areas" (those regions in commuting distance of a city of 50,000 or more), and that small center areas have increased their share of the state's total manufacturing employment by 10 percent during that time.[7]

Thus, not all small towns are in economic decline nor are they all being revitalized. Some are sharing in the nation's economic growth and prosperity while others are experiencing economic redevelopment. However, many remain in the throes of seemingly unending stagnation and decline. Two-fifths of the small cities under 10,000 report a serious loss of their central business district.[8] A third of the small cities reported that they did not have enough industry.[9]

Local Economies

The most important question to small towners, as it is to almost everyone, is how to make a living. To understand the where and how of livelihoods, the first step is to study *employment*. Perhaps more correctly stated, employment studies are conducted to understand the means of generating *income* in a community. For a small town, this body of economic data is relatively easily obtained. This requires looking for the following categories of information:

(1) what are the sources of income of the town's residents?
(2) what proportion of adult men and women are part of the labor force?
(3) in what industries do they work?
(4) what are the occupations of residents of the town?
(5) where do members of the labor force go to work?
(6) how much do the town's workers earn?

What is learned from these pieces of information?: the flow of money into the town, the proportion of people earning wages and salaries as opposed to those receiving a large portion of their incomes from welfare, social security, pensions, and other sources of governmental assistance. An appraisal of the overall level of community and per capita wealth is the result of asking these questions. Investigating employment and income provides not only a sense of how people make a living, but also a general impression of the town's standard of living. A high percentage of welfare recipients, a significant proportion of pensioners, a number of people with seasonal, low-wage, or part-time incomes—are all indicators of a local economy that is not functioning well in supporting its residents. It is the economic institutions that generally provide the means whereby individuals and families earn their living. In modern societies the economic system is extra-local since few if any communities produce only goods and service that its residents use. However, as Poplin suggests,

> It is in local communities that people find jobs, earn money, and make the majority of their purchases. This is one reason why human beings live together in communities and why they will always do so. Hence, the provision of jobs, goods and services by local businesses and industries must be thought of as one of the functional requisites of community life.[10]

In fact, Form and Miller not only perceive the community as an economic mechanism, but indicate how industry, for example, shapes the community:[11]

(1) determines location of communities;
(2) influences the size of the community;
(3) shapes the growth pattern of the community;
(4) differentiates the community by functional type and occupational composition;
(5) affects the total land-use pattern;
(6) shapes the power and social class structure of the community; and
(7) influences community character.

To study the local economy means to gain an understanding of how men choose, with or without the use of money, to employ scarce productive resources, to produce various commodities, and to distribute them for consumption—the *production, consumption,* and *distribution* of goods and services. The economy comprises many and diverse units of social organization, including retail and wholesale stores, banks, factories, labor unions, and the like.

Economic Analysis

The economy of communities has been approached in a variety of ways. Sanders suggests that there are essentially three approaches: (1) anthropologists who pick out basic characteristics and cultural themes; (2) sociologists who analyze the division of labor and occupational distribution; and (3) economists who view the economy as a major system with subsystems such as commerce, industry, organized labor, transportation and utilities, finance, and agriculture.[12]

Social scientists tend to describe the local economy. For example, Vidich and Bensman, very aware of economic forces, provide the following description of the Springdale economy:

> Next to agriculture products, lumber is the chief economic resource of the community, but it is exploited by only one commercial sawmill which is owned and operated by two families. Village business, for the most part, consists of retail establishments and service facilities. Big business is composed of several farm implement dealers and grain merchants who service the surrounding agricultural area... Economically the village functions as a farm trading center.[13]

Most observers very rarely consider the small town economy by itself. Usually they either include the small town as part of a larger area—a county, region, or metropolitan area—or interpret the small town as though it were a smaller version of a city, generally separable from the rest of society. Some economists conclude that, due to the nature of economic analysis, with its focus on systems that are open to trading goods and services with the external world, the small community has to be analyzed with the modified techniques of regional or city economy.[14] Because most data are collected for large entities, such as the county or the labor market area, some amount of local investigation and observation to fill in gaps in available information will be necessary.

Basically, the use and movement of goods and services in the small town can be divided between two sectors. The first is the private sector, comprised of those institutions and organizations that engage in production in the free market. The second is the public sector, which involves the provision of goods and services through government institutions. The latter is presented in Chapter 5. The former, the interactions of businessmen and consumers in the small town, is the focus of attention here. In the private sector, the initiative for economic activities comes from the actors involved in day-to-day decisions, basically entrepreneurs—businessmen, farmers, corporate executives, and so on—all out to enhance the profit of their firms. To understand the small town economy, the private sector may be analyzed in four segments: (1) location analysis; (2) the economic base; (3) economic function; and (4) labor market analysis.

Location Analysis

What draws industries to certain towns among the thousands of other small towns all competing for economic investments? The technique that helps explain why certain industries move to one area over another is *location analysis*. Economic firms in most cases do not locate themselves randomly. They are out to decrease costs and thus increase their potential to make a profit—firms locate where they can minimize costs and maximize profits.[15] Durr identifies three categories of industrial location:[16] (1) resource-oriented firms, (2) market-oriented firms, and (3) "footloose" industries to which nearness to resources or markets is not a major concern. Though markets and resources may be significant, if not dominant considerations, they are rarely considered in isolation. And despite appearances, many decisions to move to footloose locations are not as devoid of economic reasoning as they would seem.

A comprehensive location analysis might consider five factors as important determinants of the location of commercial and industrial firms.[17] One is the location and accessibility of *raw material* inputs. For example, many firms locate as close as possible to sources of necessary raw materials, particularly if the transportation costs of the unfinished resources are high. A second factor is *labor*: the competitiveness of labor costs and availability of labor. The nature of the supply of labor (skilled versus unskilled, old versus young) and wage rates might be taken into consideration by potential business investors. Third is the location of *markets* to which the final product is sold where raw materials are not a major concern, closeness to the intended market may be critically important. The *costs of transportation* from raw material source to processing to final market sale is a fourth factor. Many businesses attach great importance to high quality or low cost transportation facilities—good highways, major railroad linkages, and sea transport, and so on. Coordinating production and transportation as closely as possible helps businessmen eliminate costly terminal fees that are charged each time a product is transferred from one mode of transportation to another.[18] The fifth category can best be described as *"external economies"*: the location of complementary businesses close by, bonuses of low taxes and few regulations, availability of particular amenities that people value are other factors that are considered in industrial locations. Comparing these factors in a town or region against those in other areas helps to identify what economic firms are most likely to locate there.[19] To an extent, location analysis identifies the "comparative advantages" (or disadvantages) one community has over others in competing for economic investment.

The particular industries drawn to various towns differ in many ways, perhaps the most important of which is in their tendency to fluctuate. Chapin suggests five characteristics of industries that are least susceptible to sharply changing economic fortunes:[20]

(1) those industries whose products cater to a broad consuming base (e.g., low- and medium-priced products);
(2) those with products whose purchase by consumers cannot be indefinitely delayed (nondurable goods);
(3) those industries with favorable long-term growth prospects;
(4) those extractive industries with substantial resource reserves as yet untapped or replenishing;
(5) those industries offering consumer goods and services as opposed to producer goods and services.

The logic of Chapin's categories is quite clear. Producers can postpone further capital investments during economic downturns until the economy improves, but consumers can afford no such delays in the purchase of perishable or nondurable necessities. The products with wide market appeal are likely to be more stable items than those that are affordable only by those with much money to spare, who might spend that extra cash on some other frivolous product.

Some research has gone into understanding which types of firms are most likely to locate in small towns. The Stanford Research

Institute[21] has examined national data on industrial location and concluded that among the industry types that are as likely to locate in small towns as elsewhere are food and beverage products, clothing, fabricated metal products, and wood products. Those more particularly associated with small towns per se are textile industries and those closely identified with raw material availabilities, such as pulp and paper, stone, clay, and glass.

Economic Base

The *economic base* is that portion of the local economy that "is composed of the producers of goods, services, and capital who export all, or a predominant part, of their output."[22] Tweeten and Brinkman offer a rationale for the importance of this analytical approach: "The cornerstone of the export base theory is that *local markets exist only because of export markets*. ... An area grows because it earns money from goods and services supplied to other areas, and the earnings in turn provide the means to import goods and services."[23] Those industries that export their production are called "basic." The rest of the local economy, that which does not produce for export to other areas of the country, is called "nonbasic."

Based on some common measure, whether gross sales, payroll size, number of employees, tax receipts, or whatever, a ratio between the basic and nonbasic sectors can be determined. Generally, the nonbasic or residential portion is much larger than the basic sector. This ratio can be used as a multiplier to predict future growth in nonbasic activities as a result of basic production. For example, in one community, the ratio between basic and nonbasic employment may be one to three. For every job in the basic sector, such as manufacturing, there are three other residents employed in the nonbasic or service sector (such as retail shopping). Therefore, 100 new manufacturing or "basic sector" jobs can be expected to create 300 more nonbasic or service positions. In other words, assuming the town's export industry plans an increase in its operations, the basic/nonbasic ratio will help predict the benefits that will accrue to the local market sector as a result of the export sector growth.

Though economic base is a widely used concept, it has its problems. The prime difficulty is in determining just what proportion of a firm's output is "export." Common-sense notions that certain industrial classifications, such as mining and wholesaling, are "export" while retailing and government are "local," are not totally adequate. How, for example, would we automatically know how much a local manufacturer's production is export or local-market oriented? Or, what would we do about the local farms that sell part of their produce to the townspeople, and part to external markets? The small towner who tries to apply the economic base concept to his or her town will have to take advantage of the town's smallness and directly ask businessmen, farmers, and manufacturers how much of their production is export-oriented.

In many small towns, the economic base may well be nonexistent. Many small towns may only serve as bedroom communities for workers who commute as much as thirty, forty, or fifty miles each day to their jobs. The lack of basic production does not mean that no other jobs can or do exist. The town's economic survival may in large measure be due to the good fortune of its location in a healthy labor market. Therefore, in cases where the small town is analyzed, it may be wise to resort to the level of the labor market and determine a ratio between the market area's basic employment and some factor of the town's development. Of course, there are those who challenge the "causality" attributed to the economic base. Blumenfeld, for example, asserts that it is not the economic base supporting the local market sector, but just the opposite, that the local sector supports the export sector.[24] Otherwise, how else would we explain the phenomenon of a pleasant Iowa service and trade center attracting a manufacturing plant to locate there?[25] Nevertheless, without resolving which is the chicken and which is the egg among basic and nonbasic industries, the ratio they form does have some value in predicting future economic development.

Economic Functions

Economic base, employment, and location analysis all provide the basis for identifying small town functions. The particular industrial mix and locational pull or advantages of a small town or labor market combine to give each town a distinctive character. Just by common-sense observation, farmers know which towns are service and trade centers, the places where produce is sold and shipped and where equipment and supplies are purchased. Many of the small towns of central Appalachia are characterized by the dominance of mining activities. In New England, the small towns next to major rivers are known largely for their textile mills. Economists employ a similar sort of *functional analysis* to understand the distinctions among the economic purposes of communities. Why are functional town classifications important to understanding the economic dimension of small towns? Just to identify a mining town as a mining town is not very far advanced over our common-sense understanding of things. The purpose is that once we identify the function of the town, we should be able to deduce other characteristics[26]—perhaps the demographic make-up of the town or its relationship to other functional types of towns, or the probable combination of forces that caused its development, or the predisposition of residents toward particular values.

Though there is no one way to cut the pie, Harris' proposed categories of community functions have been widely accepted and highly regarded.[27] Although they are used to fit a wide range of various-sized communities, many small towns will be found in almost every category:[28]

(1) manufacturing towns
(2) retail centers
(3) diversified (multifunctional) communities
(4) wholesale centers
(5) transportation centers
(6) mining towns
(7) university towns
(8) resort and retirement towns.

Another category was added after Harris first formulated this classification in the 1940s. The process of the meteoric rise of communities with only the function of providing homes to people who worked elsewhere has been termed "suburbanization." However, not all of these "bedroom" or "satellite" communities are suburban—outside of metropolitan areas, many towns function only as residential communities. Another category worth adding is the "governmental" function which describes those small towns that are basically the seats of the county courthouse.

Once a town is classified by its function, what can be said about it? Some generalizations can be made about small town functions.

(1) Small towns known for their manufacturing are more often than not going to be predominantly one-industry towns. The size of the labor force is not sufficient to support a more complex industrial base. The single dominant industry is likely to play a major role not only in the employment of residents, but in the shaping of community values, the influencing of decisions, and the initiation and direction of community changes.[29] An important distinction in this category is whether the industry is locally controlled or absentee-owned. Mott, among others, asserts that the absentee-owned firm, affiliated as a subsidiary of a larger corporation with an out-of-town corporate headquarters, will not take a direct role in community political affairs, but will rely on public relations techniques and other indirect methods to pursue its economic self-interest.[30] A local economy's reliance on a subsidiary of a larger corporation also means that decisions on the economic fate of the community are not being made locally, but in such corporate headquarters as New York, Detroit, Pittsburgh, or Atlanta. Thus, loyalty to the particular town location is in most, though not all, cases likely to be higher on the part of the local factory owner than with the giant corporation where the town is only one of dozens of far-flung sites in its larger corporate system.

(2) Retail sales or trade communities are frequently associated with agricultural areas,[31] acting as service centers for farmers and residents to purchase equipment and goods. A reading of the economic history of small towns suggests that once-healthy agricultural trading centers, if they begin to lose that function, are "fated to become a type of industrial suburb."[32] Despite the "lamentable" shift from locally owned "Mom and Pop" stores to chain stores and supermarkets, these branch businesses have brought stability to small-town merchandising.

An important element in retail center survival is the maintenance of a people-oriented, attractive downtown. Cleanliness, a pleasant environment, and minimal auto congestion and noise are common priorities of retail sales or trade communities. Wholesale trade, on the other hand, is a function rarely dominant in small towns, the minimum size required for the wholesaling function is well out of the small town range.[33]

(3) Functionally diversified towns tend to be more stable than other, more narrowly oriented communities. In some, the variety of occupations and industries begins to resemble that of larger cities. However, where the city has a broad enough economic base to withstand economic shocks, in some small towns diversification is really a house of cards that may topple. The "interdependence" of businesses in a small town implies that a downturn or failure in one makes the economic position of the others that much weaker.[34] Diversification also implies more competition among economic institutions for land, labor, and community influence.

(4) Transportation centers can range from a railroad town to a glorified truck stop. In general, they are strongly affected by changes in transportation technology. For example, as the construction of the interstate highway system made trucking a more competitive mode of transportation, rail and shipping both suffered declines that were reflected in the positions of communities dependent on them. Closely associated with transportation communities are extractive communities such as mining towns. Extractive resource communities, ranging from coal to lumber, are generally one-industry towns, and the workers are mostly "blue collar." However, even more than in manufacturing, extractive industries are highly unionized. Small towns are as often as not relatively hostile to unionization. Sometimes unions set workers and employers at loggerheads, destroying the community feeling they should have as residents of the same town. But against domineering mine or factory owners in one-industry towns—and especially absentee owners—unions increasingly became a mechanism for workers to protect their basic interests in the work place and in the community.

(5) Open space, recreational activities, and enhancement of the environment are prime concerns in keeping small resort towns healthy. In addition, residents have to cope with the almost inevitable fact that resort economies are seasonal—small towns in New Hampshire and Vermont bustle with activity during the winter months, while the summer resorts of New York State are almost devoid of people after Labor Day. A number of other kinds of functions, such as resource extraction or heavy manufacturing, would seriously detract from the town's attractiveness as a result.

(6) Satellite communities, those characterized by a substantial proportion of the labor force working in a larger, nearby town,

find their purpose in providing a good environment in which people can reside and raise families. Along with environmental concerns, a residential community ought to be concerned with providing those services considered essential to making the town a nice place in which to live. University or college towns are known for the presence of youth and an educated, white-collar population. However, colleges are often tax exempt and thus place added tax burdens on local property owners. Government towns, like university towns, have a white-collar identity. For the most part, small towns are rarely governmental centers for political units larger than a county. At that level, particularly when the county form of government is not especially strong, the designation of a community as a county seat may not contribute much to its relative prosperity over other nearby communities. The likely benefit of the governmental function is the secondary development of a shopping and trade function to service the people who come to the county courthouse.

Labor Market Analysis

The difficulties with small town employment analysis are tied to the size of the area in question. A small town is generally not a wholly self-sufficient area. Normally, it is only part of a larger "labor market," extending beyond the borders of the small town. A low-income community, characterized by high unemployment and a relatively high dependent population, may be a weak link or insufficient beneficiary of a healthier area-wide labor market.

If the labor market is the basic unit of analysis, what does one look for in terms of evaluation? Saying, "this is a blue-collar labor market," or "this is a low-income town" is not sufficient. A blue-collar community, with a mix of industries, may well be more stable, and over the long run, better off than a white collar town with less stable employers. The low-income area may be characterized by seasonal unemployment or long structural unemployment.[35] Labor market analysis takes information on employment, occupation, incomes, community, and so on and looks at it in three ways.[36]

Industry mix. What industries operate locally and support local residents? As a rule of thumb, small town economies do not have the complexities of metropolitan areas. As it relates to small town areas, "mix" does not necessarily conform to the analytical requirements of mix in cities. What one looks for is whether employment is concentrated in one or two major industries, or whether it is scattered and not so dependent on a few employers. Are the employing industries relatively disconnected and autonomous, or are fluctuations in one felt among the others? Finally, are the types of production and commerce contained in the local labor market's industry mix stable and dependable or more subject to ups and downs?

Occupation mix. This concerns who comprises the workers in the labor market. The critical indicator is skills. A more highly skilled labor force will probably adapt to changes in the local economy with more success than will a low-skill labor force that will need more retraining. Not only is the skilled/unskilled distinction meaningful in terms of economic flexibility, it conveys a sense of income and earnings—though not necessarily. Generally, the skilled worker will earn more than the unskilled worker, the professional employee more than the technician, but not always. Still, a mix of higher skilled workers means a greater likelihood of larger incomes.

Intra-area factors. In this category is the analysis of what might happen to the labor market if it faces a major change. If a crisis hit one of the area's major employers, could the necessary interindustry transfer of workers be accomplished, or are the differences among the types of employers' needs so great that displaced workers are unable to be absorbed elsewhere in the labor market? Successful transfer of workers requires some or all of these four conditions: (1) comparable industrial or occupational lines among employers in the market; (2) comparable skill levels; (3) comparable wage rates; (4) reluctance or difficulty on the part of workers to move to other parts of the country for new jobs. In other words, the intra-area analysis considers how *autonomous*[37] a labor market is. A lack of autonomy would be suggested by likelihood of displaced workers looking to other labor markets for new positions rather than staying where they are.

Economic Development

Production, consumption, and distribution are all dynamic qualities. The economies of small towns are either heading up or down, they are growing and progressing or declining and deteriorating. Of course, the direction of a community's economy is not irreversible—once a "boom" town, in later years the community may go "bust," only to prosper during more favorable conditions. What causes one community to plunge while another maintains a healthy economy?

Small Town Decline

Dirksen, Kuehn, and Schmidt[38] suggest four alternative explanations for nonmetropolitan community decline. The first is the settlement pattern hypothesis. A community suffering decline and poverty today reflects what originally happened to it years and years ago. The initial reasons for settling in the community and the patterns of economic and land development associated with those reasons combine to limit the opportunities available to the area in the future. For example, an Appalachian community, whose primary historical event in economic development was coal mining in the 1800s, might have been so structured toward a one-industry economy that the eventual decline of the coal industry found the area without the competitive basis to lure in substitute industries.

The second is the matrix location hypothesis. Simply stated, this theory assumes that the further the community is from a metropolitan center, the less flexible the economy, and the less it is able to adjust to economic fluctuations. The nonmetropolitan "hinterland" is exploited by the urban metropolis. The income garnered from the small town's production is generally inadequate for purchasing the processed goods offered by the urban metropolis. In other words, the nonmetropolitan communities suffer a deficit in their balance of trade with metropolitan areas.

The third is the rural ghettoization thesis which asserts that three distinct processes combine to hold some nonmetropolitan communities from growing. These are: (a) "intergenerational familial poverty" (the declining community tends to be made up of families with long histories of poverty and who, as a consequence, lack initiative); (b) "class-selective migration" (trained and educated residents tend to leave the community to search for jobs elsewhere in the country, while unskilled, low-income people move in to take their place); and (c) "changes in the productivity of economic and social institutions" (public services are inadequately supplied, and low-wage private firms are all that are drawn to the area).

The fourth is economic base in reverse. Where economic base theory assumes that economic growth is some multiple of export-oriented industry, this theory suggests that decline is related to improvements in agricultural technology (decreasing the number of farm laborers needed), improvements in highways and transportation technology (increasing the possible area of shopping and jobs, thus opening up more competition to local businesses and employers), and increases in the economies of scale.

Thompson asserts that the decline of small towns, particularly those characterized by chronic unemployment and low income, can be explained by what he calls the "urban size ratchet."[39] That is, above a certain population size, growth becomes relatively self-sustaining. At least the city will not suffer the economic ups and downs that beset the smaller community. Below that critical size, economic fluctuations impact seriously on the town, causing population losses that are proportionately unmatched for metropolitan areas. As Beale says, "the faster our national population has grown, the faster and more extensively our small communities have declined."[40] In economic terms, a 45 percent increase in private nonfarm jobs in rural areas means a population increase in those areas of only 0.1 percent, but a 33 percent job gain in urban counties is matched by a population increase of 15 percent.[41]

What explains the metropolitan area's apparently greater ability to maintain economic progress? Thompson cites five factors: (1) the larger the community, the more likely it will be able to develop a diversified economic base better able to attract other industries and to withstand economic downturns. (2) Size equals political power: the larger metropolitan areas "outmuscle" small towns because of their greater electoral vote, and thus receive a disproportionate share of governmental outlays and benefits.[42] (3) With size comes investments in sewers, roads, and utilities; categories of social overhead or infrastructure whose costs are rarely absorbable except through the use of simple technologies, at the scale of the small town. (4) The lure of a nearby market, represented by the large population of the urban area, is stronger that the lure of being close to sources of raw materials and supply than the small town offers. (5) Cities are frequently more open and amenable to innovations and new ideas than are small towns. Many businessmen would rather test out a new investment scheme in the city rather than in the small town. The city is often the leader in change, followed behind somewhat reluctantly by the more traditionally oriented small town.

Economic Revitalization

The disadvantages faced by small towns due to "smallness" do not always condemn them to economic decline, instability, and low standards of living. Some small towns make progress, maintain prosperity, and provide adequate goods and services to meet the needs of their residents. What explains their apparent good fortune?

Tweeten identifies three basic economic development strategies:[43] (1) focusing on the development of basic manufacturing near appropriate markets—and assuming that service industries will follow in response; (2) training and utilizing skilled workers, recognizing that the character of the labor force has assumed greater importance than other factors of production—raw materials, capital, and so on that are more mobile (amenable to changes in location) than labor; (3) offering concessions, such as tax abatements or high-quality public services or public investments, to lure and retain private sector employers.

All three items are components of the economic planning of many communities. The objective is to lure in industry to invest in the community. The ideal of industrialization of the small town is the common cry of many local economic development advocates, particularly the Chamber of Commerce. A local economic development committee, dominated by businessmen and other town leaders, may be organized to contact potential investor firms and describe the benefits of the town. Working with local government, these local development "boosters" press for tax abatements, public improvements, and other actions which may induce investment. Boosterism for industrial development is common to small towns around the country. It is based on what Tweeten and Brinkman call the "internal combustion" theory of economic development, that the key to economic development is what communities themselves undertake to achieve it.[44]

This nation is in opposition to the idea espoused by those who believe that natural or "ecological" factors—access to raw materials, skilled labor, and markets—are the primary determinants of businesses choosing to move to small towns. A telephone survey[45] of economic development boosters in 162 small Illinois communities revealed that the respondents, who included bankers, municipal officials, planners, businessmen, and chamber of commerce officials, were split just about evenly on whether ecological factors or community action caused new firms to invest in their towns. But further analysis showed that those towns successfully achieving industrialization were not equally endowed with attractive ecological factors. Indicators of a fortunate geography—location in a metropolitan area, proximity and accesses to an interstate highway, and the like—could not be claimed by all. Even if those individuals involved in the industrialization efforts did not perceive this, other factors, such as the availability of an action plan, identified development sites, and organized groups working on attracting investment, might have made the difference between successful and unsuccessful economic development ventures, regardless of geography or ecology. The history of Appalachia, an area of plentiful resources, demonstrates that natural endowments may not always yield community prosperity without appropriate organizational and institutional capacities and actions.

As widespread as the impetus for industrialization is, its benefits are not always the same in all communities. Numerous economists substantially agree with the conclusions of Williams, Sofranko, and Root, that "[o]ften ... the impact of industry in rural communities has been shown to be illusory, negligible, or even negative. ... Still, these negative aspects of industrialization are frequently overlooked, largely because of the promise industrialization holds for improving conditions in small communities."[46] The intended benefits of industrialization are both direct—new jobs—and indirect—increased economic activity elsewhere in the community due to new investments. What economists look for are *multipliers*, much like in economic base analysis—a relationship between the new investment and the other economic activity that investment spurs. Boosterism advocates hope for a big multiplier, since the larger the multiplier, the more the community benefits. That makes sense, since the backers of boosterism—local merchants and bankers—stand to benefit from industrialization, expecting new workers to patronize their stores and shops and hoping to provide the new industry with needed supplies.

Though the entrepreneurial sector may fare well, there are problems with the whole community's ability to benefit from industrialization. Frequently, the expected multiplier is nowhere near as large as many hoped, due to what economists call *leakages*. Here are some typical examples. Commuters are one case—when the new job goes to someone from out of town, the new salary is not spent locally, but at the commuter's home. The potential economic activity from his or her salary being spent in the industry's host community "leaks" out and is lost. Similarly, the worker who used to commute out of town, and spend his or her salary back home, but now takes the new local job no longer brings in money from outside the town, thus reducing the amount of money potentially available in the town for commerce. Had another local person taken the new job while the commuter retained the out-of-town position, the external money would not have "leaked" out. Leakages also occur when local workers in existing firms switch to jobs in the new industry, but their old jobs go unfilled. As a result, the number of jobs does not increase, and the expected multiplier is reduced. Frequently, the generation of new service or support (nonbasic) jobs in response to the new industry just does not occur, leading three researchers who examined industrialization in Iowa small towns to pronounce the multiplier effect of industrialization overrated.[47]

Other costs and problems are associated with industrialization. Who gets the new jobs associated with new industries locating in small towns? Summers reports that a relatively small percentage of the new jobs is filled by the disadvantaged populations of small towns—the poor, the underemployed, and the unemployed.[48] New industry often needs skilled workers, few of whom may be located in depressed small towns. Therefore, they will import their new workers from elsewhere rather than hire locally. For the firms who are hiring less skilled workers, many will hire women who were not previously in the labor force over unemployed male heads of families.[49] The reason is cost—women just entering the market are usually lower paid than previously employed males. For women, industrialization is quite a boon, opening up alternatives to the service jobs (waitressing, housecleaning, and so on) usually reserved for women in small towns where few job opportunities exist. But the industries' ability to hire women not in the labor market may mean that high unemployment rates will not be greatly affected by the new investment. Using women for semi-skilled and unskilled positions and hiring out-of-towners for more highly skilled slots, expecially administrative and management positions, leaves many local employment problems intact.

Who pays for industrialization? Usually, it is the taxpayer. Most small towns offer some sort of tax abatements or other tax breaks to lure new businesses to invest. In addition, the municipal government will install various public improvements—sewerage, water, roads—to development sites. Since the new industry will not be paying its full tax share due to the abatement or the like, the burden for the public-provided investment incentives falls on local tax payers. Closely correlated with new investments by industry is population growth[50]—new workers moving into town and prior out-migration stemmed. Increased population means greater demand for public services—police protection, education, better roads, sewerage—by the new families in town. Many communities are surprised to discover how little slack there is in existing public works systems. Excess capacity is quickly eaten up and new roads and sewer lines must be built, teachers and law enforcement officers hired, and so forth. The payment for all this comes from tax revenues, in small towns, most often property tax revenues. With industrial investment comes increased property values. In the long run, the homeowner benefits because the property he or she owns will be worth more. In the short run, property he or she may go up along with values, meaning a heavier tax burden for the homeowner.[51]

The lesson to be learned is that industrialization, that major component of small town boosterism, is an "iffy" proposition. Payroll and employment leakages plus extra local government costs make industrial development occasionally a burden rather than a boon. Brinkman asserts that communities with much shopping and many underemployed or unemployed workers to fill new jobs may get a good multiplier from new industrial investment, perhaps as much as 1.5 to 2.5. The community with poor shopping and many commuters could see multipliers of 1.0 or lower.[52] Extending the area of analysis beyond the immediate community allows the inclusion of benefits happening outside the small town, such as the employment of other workers, thus

The Economy of the Small Town

increasing the multiplier (by "internalizing" leakages). Firms that tend to need a great deal of supplies and materials from other local businesses will also create higher multipliers. The extension of the geographic area beyond the small town and the concern for industries which are likely to become interdependent are two highlights from a more regionally oriented approach to economic development problems of nonmetropolitan small towns called *growth point* or *growth center* planning.

Actually, there is a distinction to be made between growth centers and growth points, according to Brinkman and Tweeten. They define the growth point concept as economic activity within a region clustering "around a small number of focal points." Growth center theory "emphasizes the dynamic growth of a central city and its interdependency with outlying, smaller cities and towns."[53] A common aspect of both is that they rely on *economies of size or agglomeration*,[54] a scale of economic activity which can be self-sustaining. Growth points or growth centers are now officially part of national policy for underdeveloped or lagging regions, as contained in the programs of the Economic Development Administration (EDA) and the Appalachian Regional Commission (ARC). The idea of this approach is to focus investment at a particular point—a city or group of cities—based on the hope that economic activity will spur more economic activity, promoting development over the growth area as a whole.

This presumes an ability to identify which communities, of what size, and with what natural factors are best suited to be the jumping-off points for economic growth in a region. For example, given Thompson's concept of the urban size ratchet, there may be some communities too small to be successful growth points. Some researchers suggest that a population of 20,000 is the minimum size for a growth center,[55] leaving small towns below 10,000 to receive the benefits of growth center progress rather than to become centers in their own right. However, not all communities above the urban size ratchet necessarily qualify as growth centers either. A declining city may be a poor candidate for a growth center regardless of its size. Therefore, communities with a history of a solid rate of economic growth are often selected rather than seriously lagging communities, leading to the bias that growth center policies aid those areas that are better off and largely ignore traditionally underdeveloped regions.

Another important requirement of growth point policies is the ability to identify growth industries, businesses likely to induce other economic development in their wake. Certain industries are rapid growth industries, with the ability to create many linkages to other actors in the economy. These are obviously the most attractive industries for a growth center. But interindustry linkages do not automatically occur, even among the best-suited industry. That is why the major activity of growth center policies is a huge investment in public facilities and infrastructure in and around the growth center creating an environment to make rapid growth and many interindustry linkages possible. Highways are frequently a significant part of that public investment, making isolated rural areas more accessible to major markets.

There is frequently criticism that the amount of investment in the growth center would be better spent if allocated more fairly in lesser amounts to several communities in a region. But growth point theorists defend the disproportionate allocation of resources to one point by arguing that concentrating the action there will maximize income and development over the whole region better than dispersing and dissipating resources over a larger area with diminished effect.[56] The assumption that the area surrounding the growth center receives worthwhile benefits in some proportion to the benefits received at the center itself is not fully demonstrated by the results of ARC and EDA practice. Growth center investments have tended to serve the midsized cities while offering little to the hinterland of the region. For the many small towns and hamlets surrounding the growth center, the highway investments tied to this strategy are often exits to relocate in the city rather than avenues for spin-off investments from the center. Small town residents in many areas are yet to receive the economic growth and development allegedly transmitted from their growth centers.

A final point worth noting about growth centers is the structural imbalance this strategy causes for the growth region.[57] Growth center theory comes from the concept of "growth poles," first proposed by a group of French economists to refer to central cities and their ability to induce economic development in their "zones of influence." The term pole is important because the growth center approach is based on polarizing a region between the lagging hinterland and the growing central point (or pole) dominating the region. The structural imbalance occurs regardless of whether the growth strategy is successful or unsuccessful. By definition, resources and investment are concentrated in the center, and the rest of the region, even if it benefits, will never approach the center's growth and wealth. Richardson suggests that this is inevitable.[58] These lagging areas would stagnate with even lower incomes without the investment in the growth pole, so the disparities between center and hinterland are tolerable as long as the region as a whole manages to develop economically.

Small towns, too often too small to qualify as growth centers in the minds of program managers, are frequently part of the structurally lagging hinterland in this approach to nonmetropolitan economic development. Need that always be the case? Darwent suggests that "polarized region[s] can exist at any scale, and smaller ones, polarized around smaller centers, will tend to 'nest' within larger ones."[59] Therefore, small towns below 10,000 might be able to serve as growth centers for the hamlets and villages surrounding them, transmitting economic benefits from their limited economic activities to their smaller hinterland. Thompson suggests that if small towns are not of the size or scale to generate self-sustaining economic activity and growth, several small towns linked by good transportation might together form a loosely associated labor market, and as such achieve the scale and diversification necessary to overcome their individual economic shortcomings.[60]

Conclusion

Economic development strategies stressing local industrialization boosterism or nonmetropolitan regional growth center strategies are not "tried and true" remedies which work everywhere. With the former, it can be shown that industrialization imposes many costs on a community, often in excess of the benefits it brings. Who bears the burden of those costs? Who receives the benefits? New industrial development may boost a town's tax base and increase local payrolls, indicators of *community* prosperity. These same economic improvements, however, may totally ignore the plight of the town's poorest residents, its low-income or welfare population. In some instances, to make room for the employees of recently located firms, the homes of low-income families are razed and replaced by suburban-style subdivisions for the families of new residents. The displaced residents often do not find alternative housing within the town's boundaries and are forced to look elsewhere. The community's new prosperity is not shared by those residents displaced or ignored. Not every one benefits equally from community economic development. It is important therefore to exercise caution in selecting measures of success and prosperity. What might be good progress for the *community* (a high multiplier, for example) might not help or might even adversely affect many individuals.

This issue is commonly referred to as *equity*—fair treatment for all groups of people in the community. If costs are to be borne, everyone in the town should bear them in proportion to their capacity; if public benefits are to be received, they should help those most in need as well as the general population. One reason small towns often overlook the equity issue in economic development is in part due to the ignorance of great disparities of wealth and resources in small towns. Despite the image of middle class homogeneity, small towns are characterized by great gaps between the wealthiest and poorest residents. Boosterism is largely the province of the bankers, businessmen, and professionals in small towns—by national standards, they may not be very wealthy, but in small towns they generally constitute the upper crust in economic terms. That they might benefit from local boosterism is not surprising. Neither is the invisibility of the small town poor. According to Clavel and Goldsmith, much nonmetropolitan poverty is "hidden."

> The dispersion of the essentially urban middle class, mixed with indigenous poor populations, raises income statistics high enough so that even township level data often become too coarse a sieve to satisfactorily identify poverty, even in the income sense. The poor live scattered through towns, or clustered in rather small pockets that are hard to find, should anyone even look.[61]

Downtown improvements may enhance the attractiveness of shops and stores located there, but leave unattended the perhaps more serious problems of poverty among "reliefers," retirees, "shack people," and other low-income groups in town. An equitable small town economic development program should give adequate attention to the interests of all resident groups, including those least likely to clearly articulate their concerns but perhaps most seriously in need.

The growth center policies suggest a certain futility to economic development strategies at the small town level. Whether the guiding concept is growth center or Thompson's multicommunity labor market, the central notion is that the small town is too small to be a viable economic unit. That is not to say that larger areas, cities, for example, are fully self-sufficient. While no place is ever totally self-sufficient, the point of many economists is that small towns by themselves do not possess the economies of scale to be individually significant. They are too small to provide much of a range of goods and services to their residents, many limited to a few groceries, taverns, luncheonettes, and perhaps a few additional shops and service stores. They are too small to generate the level of economic activity, especially retail sales, to attract many commercial ventures, those stores preferring the central locations of shopping centers adjacent to major highways over the less accessible village downtowns. Small town economies are rarely diversified enough to attract industrial investors that require a significant number of linkages and transactions with other locally based businesses, leaving small towns to be one-factory or two-factory communities in the main.

The literature covering economic development planning at the level of the small town is rather limited in scope. Small businesses and industrial development may provide economic relief in some situations, but they are not general solutions. Growth center strategies and their various cousins, however planned, mandate that small towns may have to address the problems of economic development jointly, regionally. Intercommunity cooperation and regional economic planning, if implemented sensitive to the specific character and needs of the participant communities, can help eliminate demeaning and debilitating competition among small towns for a limited pool of industrial investors, the costs of attracting them perhaps outrunning the potential rewards. Regional approaches may achieve economies of scale and perhaps stability and diversification which small towns on their own are hard pressed to do.

DISCUSSION GUIDE

I. What are the relevant indicators of employment and sources of income in your town?

Using publicly available U.S. census data from the volume entitled *Social and Economic Characteristics*, identify the following conditions:

(A) What proportion of the town's adult population is in the labor force?
   U.S. (1970): 61.0%   Community: _____

(B) In what occupations do most residents work?

	U.S. 1970 %	Community %
Blue Collar:	49	_____
White Collar:	51	_____

(C) In what industries are most residents employed?

	U.S. 1975 %	Community %
Manufacturing:	22.9	_____
Mining:	0.9	_____
Construction:	5.9	_____
Transportation & Public Utilities:	6.63	_____
Retail Trade:	16.67	_____
Wholesale Trade:	3.93	_____
Medical/Health Services:	6.92	_____
Finance, Insurance, Real Estate:	5.05	_____
Entertainment/Recreation:	1.0	_____
Public Administration:	5.6	_____

Professional Services:	*12.39*	_____
Misc. Business Services:	*3.25*	_____
Personal Services:	*4.38*	_____
Agriculture & Forestries:	*4.09*	_____

(D) What proportion of the town's households rely on welfare, social security, pensions, or other "dependent" sources of income?

	U.S. (1970) %	Community %
Population receiving AFDC*:	5.3	_____
Population receiving Social Security benefits:	13.5	_____
Population receiving Old Age Assistance (OAA)**:	0.9	_____
Population receiving Aid to Permanently and Totally Disabled (APTD)**:	0.6	_____

(E) What are the median family and unrelated individuals' income levels in the town?

	U.S. (1970)	Community
Family:	$9,867	_____
Unrelated Individual:	$3,137	_____

The types of employment will hint at the narrowness or diversity of the economy—"the industry mix." The classification of jobs and sources of income reveals the skill levels and training of the labor force—"the occupational mix." To understand intra-area factors, these data categories would

---

NOTES:
*Aid to Families with Dependent Children
**These programs have been altered under the new "SSI" legislation. One should check with the county public assistance board to break down SSI recipients according to OAA and APTD categories.

have to be obtained for neighboring municipalities (which can be done without too much difficulty at the county seat) and compared for similarity of occupations, industries, and income levels. This provides a comparative sense of stability of the economy. Lacking diversity, skills, employable residents, and integration with other communities, the town's economy may be quite precarious.

II. What factors have contributed to the location of the economic activity in the local community or area?

   (A) What and where are the raw materials used in the town's activities drawn from?
   _____
   _____

   (B) What are the wage rates and skills of the labor force that make it attractive to industry?
   Wage rates:_____
   Skills:_____

   (C) Where is the market for the products of the town's businesses located?
   _____

   (D) What are the availability and costs of transportation?
   Availability:_____
   Cost:_____

   (E) What other factors, such as complementary industries, local taxes, regulations, and so on, are considerations in the decisions of local business?
   _____
   _____
   _____

   (F) Overall, which factors best account for the development of the town as it is today?
   _____
   _____

III. What is the town's economic base?

What is necessary is a visit to the town's local employers. Ask them whether their production serves markets outside the community (or local

The Economy of the Small Town

labor market), or is intended for local consumption. Add up all the workers involved in "export" or "basic" activities, compare that figure to the rest of the town's employment:

$$\frac{\text{Basic Employment}}{\text{Nonbasic Employment}} = \text{Employment Ratio}$$

Similarly, determine the payroll of the basic sector compared to the nonbasic sector:

$$\frac{\text{Basic Sector Payroll}}{\text{Nonbasic Payroll}} = \text{Payroll Ratio}$$

Or determine at the town hall the tax receipts from the town's basic employers, and calculate the tax revenues garnered from other local sources:

$$\frac{\text{Basic Industry Receipts}}{\text{All Other Local Revenues}} = \text{Revenue Ratio}$$

Or make a ratio between basic employment and total population or households:

$$\frac{\text{Basic Employment}}{\text{Total Population}} = \text{Population Ratio}$$

Community Ratios
Employment Ratio _____
Payroll Ratio _____
Revenue Ratio _____
Population Ratio _____

It may well be that the small town is not a large enough area about which to make valid predictions, so a shift to the level of the labor market may be necessary. What the economic base ratio does is allow one to predict, not infallibly however, some of the possible growth (or decline) that might occur with changes in the basic sector of the economy. That does not mean that the base causes local sector growth, just that according to theory, a predictive relationship exists. For example, if the employment ratio is 3, one would predict that nine new local nonbasic sector jobs would follow for every three basic jobs.

IV. What economic function does the town perform that differentiates it from other small towns?

It is important to understand what the main economic function of a town is—for instance, is it a manufacturing or extractive type of community? One way to determine its classification is to look at the percentages of employees in specific industries. The following are definitions based on Harris' classification scheme:

(A) *Manufacturing towns:* At a minimum, at least 30% to 45% of the workers should hold manufacturing-related occupations;
(B) *Wholesale trade towns:* Employment in wholesale trade is at least 20% of total employment, or almost 50% of retailing;
(C) *Transportation communities:* At least 10% of all employed workers are in jobs connected with transportation;
(D) *Retail towns:* Employment in retailing is at least half of total employment;
(E) *Extractive towns:* Employment in resource extraction (mining, lumbering, and so on) is at least 15% of total employment;
(F) *University towns:* One-fourth of the town's population is enrolled in postsecondary education;
(G) *Governmental towns:* Those designated as county seats, which accounts for the development of other industry (usually retail) in the community;
(H) *Resort communities:* Local industry composed mostly of hotels and motels, souvenir shops, and other tourist or seasonal activities;
(I) *Diversified:* Those with a mixture of occupational and industrial categories, in which no one category dominates;
(J) *Satellites:* One in which a substantial majority commute to neighboring employment centers.

How would you classify your town?_____

NOTES

1. Earl O. Heady, "New Priorities," in Larry R. Whiting, ed., *Rural Development: Research Priorities* (Ames: Iowa State University Press, 1973), p. 102.

The Economy of the Small Town 151

2. Charles C. Jones, *The Dead Towns of Georgia* (Savannah: Morning News Steam Printing House, 1874).

3. William Simon and John H. Gagnon, "The Decline and Fall of the Small Town," in Robert Mills French, ed., *The Community* (Berkeley: F.E. Peacock, 1969), pp. 509-510.

4. Bert N. Adams, "The Small Trade Center: Process and Perceptions of Growth or Decline," in French, p. 484.

5. E.F. Schumacher, *Small Is Beautiful* (New York: Harper & Row, 1973), p. 63.

6. E.J. Mishan, *Technology and Growth* (New York: Praeger, 1969), pp. 57-62.

7. T.E. Fuller, N.B. Gingrich, and J.D. Jansma, *Industrial Growth for Rural Communities* (University Park: Pennsylvania State University Press, 1973), p. 16.

8. *Municipal Year Book of 1968* (Washington, D.C.: The International City Management Association, 1968), p. 148.

9. Michael J. Murphy, *Governmental Data in Municipalities 25,000 and Under* (Washington, D.C.: Urban Data Service Reports, International City Management Association, 1975), p. 15.

10. Dennis E. Poplin, *Communities* (New York: MacMillan, 1972), pp. 171-172.

11. William H. Form and Delbert C. Miller, *Industry, Labor and Community* (New York: Harper & Row, 1960), pp. 18-49.

12. Irwin T. Sanders, *The Community* (New York: Ronald Press, 1958), pp. 225-227.

13. Arthur J. Vidich and Joseph Bensman, *Small Town in Mass Society* (Princeton: Princeton University Press, 1968), p. 19.

14. Richard B. Andrews, "Economic Studies," in William I. Goodman and Eric C. Freund, eds., *Principles and Practices of Urban Planning* (International City Management Association, 1968), p. 83.

15. Brian Goodall, *The Economics of Urban Areas* (New York: Pergamon, 1972), p. 13.

16. Fred Durr, *The Urban Economy* (Scranton: International Textbook Company, 1971), p. 23.

17. Steve Carter, Kendall Bert, and Peter Nobert, "Controlling Growth: A Challenge for Local Government," in *The Municipal Yearbook 1974* (Washington, D.C.: International City Management Association, 1974), p. 166.

18. Goodall, 1972, p. 124.

19. Brian J.L. Berry, "Theories of Urban and Regional Growth," in Brian J.L. Berry and Frank E. Horton, eds., *Geographic Perspectives on Urban Systems* (Englewood Cliffs, N.J.: Prentice Hall, 1970), p. 96.

20. F. Stuart Chapin, *Urban Land Use Planning* (Urbana: University of Illinois Press, 1965), p. 150.

21. R.M. Morse, O.P. Mathur, and M.C.K. Swamy, "Manufacturing Activities Related to City Size," in Berry and Horton, 1970, p. 140.

22. Andrews, 1960, p. 79.

23. Luther Tweeten and George L. Brinkman, *Micropolitan Development: Theory and Practice of Greater Rural Economic Development* (Ames: Iowa State University Press, 1976), p. 328.

24. Hans Blumenfeld, "The Economic Base of the Metropolis," *Journal of the American Institute of Planners* (Fall 1955), pp. 114-132.

25. Wilbur Thompson, *A Preface to Urban Economics* (Baltimore: Johns Hopkins Press, 1965), p. 30.

26. Robert T. Smith, "Functional Town Classification," in Berry and Horton, 1970, p. 111.

27. Chauncy D. Harris, "A Functional Classification of Cities in the United States," *The Geographical Review* (1943), pp. 86-89.
28. Harris divided manufacturing towns into two groups, the distinction between them not all that useful for small towns.
29. Form and Miller, 1960, p. 439.
30. Paul E. Mott, "The Role of the Absentee Owner Corporation in the Changing Community," in Michael Aiken and Paul E. Mott, eds., *The Structure of Community Power* (New York: Random House, 1970), p. 178.
31. Gerald A. Doeksen, John Kuehn, and Joseph Schmidt, "Consequences of Decline and Community Economic Adjustment to It," in Whiting, ed., *Communities Left Behind* (Ames: Iowa State University Press, 1974), p. 38.
32. Page Smith, *As a City Upon a Hill: The Town in American History*, (Cambridge: MIT Press, 1966), p. 107.
33. Ogburn and Duncan cite a minimum threshold population of at least 200,000 in Goodall, 1972, p. 32.
34. Doeksen, Kuehn, and Schmidt, 1974, p. 38.
35. Thompson, 1965, pp. 204-205.
36. Ibid., pp. 67-73.
37. Ibid., p. 71.
38. Doeksen, Kuehn, and Schmidt, 1974, p. 36.
39. Thompson, 1965, pp. 21-24.
40. Calvin L. Beale, "Quantitative Dimensions of Decline and Hability Among Rural Communities," in Whiting, 1974, p. 3.
41. Ibid., pp. 14-15.
42. Tweeten and Brinkman, 1976, pp. 16-18.
43. Luther Tweeten, 1974, in Whiting, pp. 91-107.
44. Tweeten and Brinkman, 1976, p. 75.
45. James Williams, Andrew Sofranko, and Brenda Root, "Change Agents and Industrial Development in Small Towns: Will Social Action Have Any Impact?" *Journal of the Community Development Society* (Spring 1977), pp. 25-27.
46. Ibid., p. 20.
47. David L. Rogers, Willis Goudy, and Robert O. Richards, "Impacts of Industrialization on Employment and Occupational Structures," *Journal of the Community Development Society* (Spring 1976), p. 57.
48. Gene F. Summers, "Small Towns Beware: Industry Can Be Costly," *Planning* (May 1976), p. 20.
49. Rogers, Goudy, and Richards, 1976, p. 54.
50. Summers, 1976, p. 21.
51. Ibid., p. 20.
52. George Brinkman, "Effects of Industrializing Small Communities, *Journal of the Community Development Society* (Spring 1973), p. 73.
53. Tweeten and Brinkman, 1976, pp. 68-69.
54. Harry W. Richardson, *Regional Economics* (New York: Praeger Publishers, 1969), p. 416.
55. Tweeten and Brinkman, 1976, pp. 435.
56. Richardson, 1969, p. 424.
57. Ibid., p. 424.
58. Ibid., p. 426.
59. D.F. Darwent, "Growth Poles and Growth Centers in Regional Planning: A Review," in John Friedmann and William Alouso, eds., *Regional Policy: Readings in Theory and Applications* (Cambridge: MIT Press, 1975), p. 548.

60. Thompson, 1965, pp. 33-35.
61. Pierre Clavel and William W. Goldsmith, "Non-Metropolitan Poverty and Community Institutions," *Journal of the Community Development Society* (Fall 1973), p. 83.

Chapter 5

PATTERNS OF INFLUENCE AND DECISION-MAKING IN SMALL TOWNS

Democracy has come to mean self-government. Lincoln in his Gettysburg Address provides a guiding American principle: "government of the people, by the people, and for the people." In other words, people should govern themselves and participate in those decisions which affect their lives. Yet, most people are governed by others. They do not participate in key community decisions but acquiesce to the decisions of others. Only 54 percent of those eligible to vote, or 67 percent of those registered to vote, actually did so in the 1976 presidential election. Voter turnout for local elections is alarmingly lower. Those who take the time to participate are (statistically) individuals of higher income, occupation, and education. Many citizens do not attend meetings to discuss community problems nor do many attempt to discuss with others matters concerning what should be done to improve their town or how their tax money should be spent. Even fewer citizens actively campaign for candidates. A small handful of people openly demonstrate and petition their government for a redress of grievances. Thus as the scope and size of government increases, the role and influence of individual citizens seems to decrease.

At the outset, we should distinguish between direct and indirect democracy. The former involves first-hand citizen participation in public deliberations, voting on issues referred to the people, or seeking recall of public officials who are believed to have violated the public trust. Direct methods require that almost all decisions are made by the people. Indirect democracy, instead, utilizes elected public officials to be representatives *of* the people and to act *for* the people. In small towns, and especially in the town halls of New England, citizens are more likely to participate directly, and thus truly govern themselves. On the other hand, in larger cities citizens rely almost exclusively upon elected and appointed public officials. There is then the tentative conclusion that the smaller the place, the more likely is direct democracy.

When we think of local government and politics, we tend to think of smoke-filled rooms where corrupt politicians play favorites or where civic leaders meet in exclusive clubs to look out for their own interests and feather their own nests. The American politician is seen slapping people on the back, shaking hands, and kissing babies. Yet, civics books have taught us to expect that concerned citizens go to the polls and vote for leaders who have the best interests of the community at heart.

Throughout history there has been a deep fascination about political, economic, and social power. Despite democratic notions to the contrary, most studies reveal that power is unequally held or is shared by only a few persons, a small group, a few institutions. Simply put, these few exercise more power and influence than others. Some persons apparently are more eager to acquire and use power, while others find it easier to follow the leader than to be the leader. Those with authority and influence may seek to further their own interests rather than look out for the interests of others or the common good. Those with influence use many methods to achieve their desired ends. Some use persuasion and reason so that others will understand what should and must be done. Some may exploit the weaknesses of others. Some apply force or penalize those who oppose them. Very often it is difficult to discern the specific manner in which those with authority and influence attempt to get their way. Of course, some persons have no power and choose either to accept their lot or to complain, and rarely do much to change their condition.

Power has been studied for a long time by analysts who have tried to discover exactly who governs, controls, or decides what is or is not done in community affairs. Others, more philosophic, have been concerned about what should or should not be done. While a community is inhabited by individuals, it also has clusters of people who collectively make decisions in an attempt to direct or redirect the course of events which determines "who gets what, when, and how." Some commentators, such as Weber, define power individualistically: "Power is the probability that one actor within a social relationship will be in a position to carry out his own will despite resistance, regardless of the basis on which this probability rests."[1] Individuals, however, often find it more useful to join with others to maximize their power and even try to monopolize it.

Social scientists have long studied various governmental forms and the decision-making process. To better understand how decisions are made in small towns we shall discuss decision-making forms and processes, influence patterns, citizen participation, the mobilization of community resources as well as public finance, and the scope of government.

Decision-Making Centers

There are formal—legal and voluntary—community decision-making centers within small towns. These centers formulate the preferences of specific individuals and organizations and then attempt to influence community policy in desired directions. They can be found at city hall, the fire hall, the country club and so on. One important center is, of course, local government. Auxiliary to local government are the political parties which select candidates, support their campaigns, shape public opinion which eventually is expressed in conversations among neighbors, friends, and family, and/or at the ballot box. Another set of centers can be found in community organizations where members and civic leaders meet to discuss their common problems and preferences. Social clubs and religious associations comprise other sets of decision-making centers.

Government

The organization of public authority at the local level involves both areal (spatial) and capital (separation of power) distribution of authority. Areal distribution of authority involves the layers or levels of governments from the national to state and local (the local level includes county, municipal, townships, and special districts). The American federal system is one of the most decentralized governments in the world. Thus, a single resident will be a member of several legal jurisdictions and be able to participate in and be served by each. In 1977 there were 3,042 counties, 18,862 municipalities, 16,822 townships, 15,174 school districts, and 25,962 special districts operating at the local level. Most of the municipalities and townships are small in size—under 10,000 in population. They comprise some 90 percent of the places but serve only 23 percent of the U.S. population.

Many small places are referred to loosely as villages, towns, and boroughs. For example, the term "village" is sometimes applied to any hamlet whether or not it is legally incorporated, while the "towns" of New England are not incorporated. State constitutions specify which term is to be used and generally use size of place to classify the type of town or city. Furthermore, they specify the authority each type of local government is to exercise. That is, large municipalities have been granted the most authority to provide more public services, and small towns or townships the least. Those places that have not been legally incorporated generally rely on other governments—counties and special districts—for local services, if any are provided at all.

In addition, much of the concern of political scientists has focused on capital distribution of authority or the separation of powers in respect to the particular form of local government. They believe that the structure of government and the procedural rules of decision-making play a significant role in community affairs. They have identified five prevailing forms of local government. The first is the *weak mayor* where the mayor has little appointive power and little to do with shaping the local budget. This is a popular form in small towns (see A in Figure 5.1). The *strong mayor* by contrast appoints departmental personnel, shapes the budget, and vetoes city council legislation. This is the prevailing form for big cities (see B in Figure 5.1). These two forms follow the national pattern of separating power into executive power and legislative power. In the strong mayor form of government the citizens generally look to the mayor to get things done and hold him or her responsible. The *council manager* form of local government is the most popular among moderate-size cities—25,000 to 500,000. Within this form, the city council, generally elected on a nonpartisan basis, appoints a general manager to prepare the budget, and to hire and supervise personnel in the day-to-day operation of the city (see C in Figure 5.1). The city-manager is expected to bring business-like efficiency to local government and reduce corruption and waste. The council makes policy and the manager implements it while the mayor is primarily a symbolic leader.

There are two forms of local government used in very small towns. The first is the town meeting which historically has appeared in New England where citizens assemble to discuss and vote

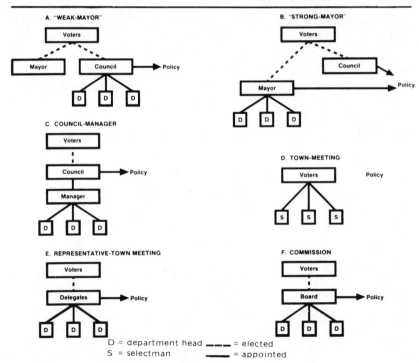

SOURCE: Adapted from R. Eric Weise and Alfred de Grazia, *Eight Branches* (Collegiate Publisher, 1975, p. 307).

Figure 5.1: Forms of Local Government

directly on community affairs (see D in Figure 5.1). The other is the representative town meeting where delegates are elected to formulate town policy and administrative responsibility for specific government activities (see E, Figure 5.1). Finally, the commission form is widely used for governing counties, townships, and special districts (see F, Figure 5.1). The elected commissioners serve as both legislators and executives as they both formulate policy and assume specific administrative responsibility.

The town hall meeting is the most accessible to citizens, and the council-manager form is the least. The latter is considered the most capable of handling routine matters, whereas a mayor-council system is considered the least efficient. On controversial issues the mayor-council form is more inclined to hear all sides and to negotiate agreeable solutions among the contestants than is the council-manager form. Thus, city-manager type governments will focus on routine decisions, while the mayor-council form makes more nonroutine decisions. Of course, in very small towns city managers are less professionally trained than those found in larger cities, and therefore may not be as effective. Similarly, small town mayors who are part-time and unpaid may not be interested in nor concerned with negotiating differences.

A general reform movement has taken place in American local government. This movement supports not only the city manager form of government but also nonpartisan elections, the "at-large" rather than the ward (section of community) basis for electing city council members, and the "short ballot" with few elected officials in city government. The reformers' thrust is to deemphasize partisan politics in local government. They believe that removing the political parties not only reduces the corrupt and ineffective conduct of local affairs, but also enhances the opportunity for key community organizations—especially business or merchant-oriented groups—to play a greater role in community affairs. In small towns there is likely to be little difference between those who dominate the local economy and those who dominate local government. They may be the same persons or, if different people, they may be generally in agreement on the key community problems, issues, and policies.

Community Organization

Hillary cautions us to carefully distinguish between such formal organizations as governments, and those community organizations that are voluntarily formed by people who join together. The voluntary organizations promote the general goals and aspirations of select groups. They advance their group's preferences or protect their interests within the locality.[2] The organizations tend to be self-selective as to their membership. They may be affiliated with national organizations such as Rotary, Lions, and the like or locally based to meet specific needs such as book, garden, and card clubs.[3] While small communities have fewer organizations than do larger cities, they seem to play a more significant role in community affairs and often serve as genuine alternatives to local governments. The voluntary fire company is a prime example. Its actions are voluntary and its projects are not funded from the public sources. When local government is reluctant to act on an issue, these organizations often can be mobilized to take action.

Those who participate in voluntary organizations tend to be middle-aged, married with children, upwardly mobile, residentially stable, and of higher socioeconomic status. Members participate more when the association is not large, not specialized, and does not have a large staff. They also participate more when the members have considerable power relative to the officers and when the membership is homogeneous in character. Berelson and Steiner indicate the broad effect of voluntary associations to be:

> (a) to promote participation in the social life, by providing frequent and attractive opportunities for personal contacts; (b) to increase personal freedom, by increasing the choice of appropriate modes of behavior; (c) to promote social change, by increasing pressures from particular organized segments of the society; (d) both to sharpen and to soften class distinctions, depending on the extent to which the associations are limited to "one's own kind" as against bringing diverse groups together in a Common Cause.[4]

Citizen Participation

Citizens participate in community affairs in a variety of ways—politically, socially, and/or economically—and to varying degrees with different motives and consequences that reflect self and/or public interests. Some persons are apathetic and uninvolved while others engage in the discussion of and the voting on many community issues. Some people attempt to persuade others to agree with them. Some want to be leaders or at least have access to or direct contact with community leaders so they might tell them what they want or what they think should be done. Social scientists have studied citizen participation as an indicator of the degree of democracy in a community or to learn what can be done to raise the level of participation in an attempt to make the country more democratic. One cardinal principle followed is that "the greater the participation the more democracy."

Verba and Nie in an extensive study of who participates, how they do so, and with what consequences, found a wide range of activities engaged in by Americans. Table 5.1 illustrates that voting is the most frequent activity and joining a political club or organization is the least frequent activity engaged in by Americans.

Verba and Nie have taken the activities shown in Table 5.1 and classified the types of participators of a sample of Americans. The most active, *The complete activists* (11 percent) are upper status individuals who come disproportionately from small towns and who are middle-aged. The *campaigners* (15 percent) are more informed and feel more efficacious than the average citizen. They have strong partisan (political) affiliation and like to take sides in community conflicts and have relatively extreme issue positions. They are overrepresented among the upper status, males, Catholics, and those that live in cities. The *communalists* (20 percent)

Table 5.1: Citizen Participation Activities [5]

Percent	Activity
72	Vote regularly in presidential elections
47	Vote always in local elections
32	Active in an organization involved in community problems
30	Work with others trying to solve community problems
28	Attempt to persuade others to vote
26	Work for a party or candidate during election
20	Contact local government official about issue or problem
19	Attend political meeting or rally
18	Contact state or national government official about a problem or issue
14	Form group or organization to solve local community problem
13	Give money to party or candidate during election
8	Member of political club or organization

combine the willingness to be quite active in community affairs while staying out of the relatively conflictful action of campaigning. They are well above average in their sense of efficacy, level of information, and sense of contributing to the community. They are more likely to reside in rural areas and suburbs, and to be upper status. *The parochial paritcipants* (4 percent) make particularized contact with community leaders. Their activity requires initiative, but their interest is very much limited to ways that will enhance their own personal lives. They are overrepresented among the young, Catholics, and suburbanities. *The voting specialists* (21 percent) seldom miss voting but do not go beyond this singular act. This activity requires little initiative and lower than average psychological involvement. The voting specialists have less efficacy, less than average information, and experience less sense of contributing to the community. They are higher than average in partisanship, habitually attached to their political party, but are less likely to take sides in community issues. Finally, *the inactives* (22 percent) do not participate at all. They are not psychologically involved in politics. They not only lack civic expertise but they are also relatively indifferent to political conflict and have little sense of the need for civic contribution. "The passive citizen, thus, comes disproportionately from these groups: those with lower social status, blacks, the young, and to a lesser extent, women, Protestants, and those who live in small towns."[6]

Voting for president is clearly the customary form of participation. While this is a common characteristic of all voters, there are important differences among them. The Bureau of the Census, which monitors elections, has found that[7]

(1) males vote more often than females;
(2) whites vote more than Negroes;
(3) middle-aged people vote more often than the young or the very old;
(4) Northerners and Westerners vote more often than Southerners;
(5) well-educated persons (high school and above) vote more often than those with grade school or less education;
(6) employed persons vote more than the unemployed persons;
(7) people with higher incomes ($7,500 or more) vote more than those with lower incomes.

By means of voting people choose public officials who they expect will represent their interests. These officials then formulate public policy and see to it that their decisions are implemented to the satisfaction of their constituencies. The reasons given for why citizens participate include: the search for information, the enhancement of one's own political career, seeking and protecting one's interest, promoting the general welfare of the community. The consequences of citizen participation may satisfy some of the people some of the time. If some individuals and groups are not pleased with what is happening, they may be drawn into community issues to express their preferences. Although a high rate of citizen participation is considered desirable by some to make democracy work, it has also been associated with increased community controversy, and thus considered by others as undesirable. If increased participation leads to controversy, then some believe community leaders do in fact respond to those who shout the loudest.

Patterns of Leadership and Influence

High rates of participation are related not only to the degree of democracy but also to certain patterns of who governs small towns. Two distinctly different influence patterns have been discovered in studies of several hundred American communities. The first, referred to as an *elite* pattern, is characterized by few leaders who rule, dominate, or control community affairs. (Other terms used for this pattern are: monolithic, pyramidal, integrated, stable, highly structured, and centralized.) The second, referred to as a *pluralistic* pattern, is characterized by competitive sets of leaders who attempt to govern, influence, persuade, or bargain to affect specific community policies or decisions. (Other terms used for this pattern are: diffused, amorphous, fluid group alliances, unconcentrated, ecology of games, and decentralized.) The elite pattern is believed to be the prevailing one in small communities while the pluralistic pattern prevails in larger cities. However, neither is absolutely predictable by size as some small towns have been found with pluralistic leadership and some large cities with elite patterns. Some communities may shift from one pattern to another, especially in times of crises or when leadership changes.

These influence patterns are not easy to see, hear, touch, taste, or smell. Part of the pattern may be seen in the leaders who preside at meetings of community organizations and another part by who is in public office. Other parts of the influence pattern may be discerned by asking community leaders who they think is influential, or from rumors about who has influence. However, none of these methods seems to provide sufficient evidence to identify the complete influence pattern.

Clark has actually sketched two types of elites and two types of pluralistic or four alternative community decision-making systems.[8] To help visualize the difference between the two patterns of influence, while running the risk of oversimplification, we offer Figure 5.2 to emphasize their contrasts. The *monolithic* or consensual elite system has few leaders on top of the hierarchy who generally dominate community affairs. There is considerable agreement among leaders on what should be done and how it should be done. This pattern is most commonly found in small towns. The *polylithic* or competitive elite system mainly found in big cities indicates elites have specialized spheres of influence. Different leaders or groups dominate governmental, economic, and cultural affairs. This system tends to introduce some competition or challenge and precludes any one person or group from dominating the community as a whole. One pluralistic system is the *competitive mass* system where leaders are not only specialized into spheres of influence, but they share power with many persons. In this case, the rank and file citizens have greater access and opportunity to make their will known and have it implemented. This leadership pattern includes representatives from many sectors of town—government officials, economic dominants, and civic leaders—who compete, bargain, and cooperate. In this situation, overall community direction and guidance comes by adjustment and compromise. The other pluralistic system is the *consensual mass* system, where there is virtually no difference between leaders and citizens because influence is not specialized and citizens participate more directly.

Three approaches have been used to determine who has community influence. They identify (1) those who control or exercise influence by virtue of their formal position in community institutions; (2) those who have a reputation for power and dominance in community affairs; and (3) those who decide by actually

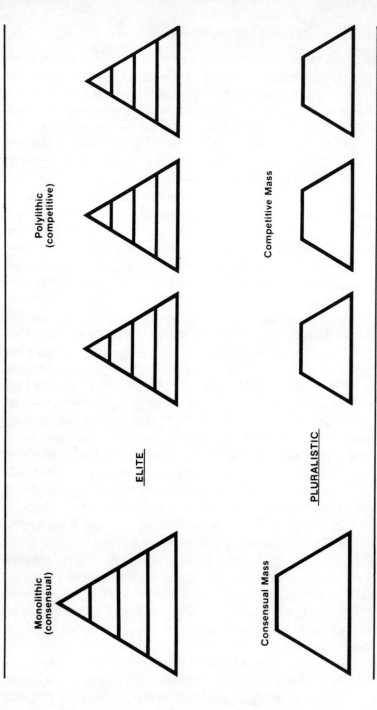

Figure 5.2: Types of Influence Patterns

exercising their influence on specific policies and issues. Any one method, briefly described, may provide some idea of who most of the leaders are and how they get things done.

The *positional* approach identifies those who are elected and appointed to be leaders in community institutions and organizations.[9] It assumes that people who hold formal positions can and do use their offices to determine what is done in the community. Leaders use their authority if they are in government, or they use their influence if they are in a voluntary organization. The governmental positions include the mayor, city or town council members, and other local special authorities. Other positions include owners and managers of local economic firms, the heads of voluntary community organizations, such as volunteer fire associations, political parties, veteran groups, luncheon, garden, historical and leisure time clubs, and leaders in local churches and fraternal organizations. An inventory of those who hold such positions in town can easily be secured first by obtaining a list of public, private, and voluntary organizations, and then by compiling a list of the officers and members of the boards of directors, trustees, and so on of those organizations. If from the list it emerges that some people hold positions in two or more organizations (overlapping positions), it is likely that an elite pattern exists because only a few individuals control many key positions. Conversely, if it is rare that one person holds two or more positions, then it is more likely that a pluralistic pattern exists. Thus, if the mayor is the head of the volunteer fire company and also the leading merchant in town, he may be the key leader of an elite system. In such a system there is probably agreement among the leadership on town matters. However, if these positions are held by three different persons, the resulting pluralistic system would probably produce some differences of opinion on selected policies.

The reputational approach first identifies a group of "knowledgeable" persons, those most likely to know what is going on and who are influencing community affairs.[10] These people, often referred to by social scientists as "key informants," may be the local newspaper publisher or editor, the mayor or city manager, a banker, a realtor, an owner or manager of a key firm, or some known critic. The questions generally asked of these people are: (1) who do you regard as the most influential people in town? (2) if a project were before the community that required a decision by a group of leaders (leaders that nearly everyone would accept) which people would you choose, regardless of whether you know them or agree with them? (3) whom would you go to for advice to get support for a specific proposal and to see that it is implemented? The answers to these questions should provide a list of reputed community leaders. Then these leaders should be interviewed for their perception of community problems and solutions as well as who else they believe to be influential. One could, of course, stop with the interviews of "knowledgeables," but it is important to secure the insights of those more involved and influential in community decision-making. In any case, those leaders receiving the most nominations are considered the most influential. If there are only a few (2 or 3) people with many nominations, the leadership is more likely to be an elite. If these few influentials belong to the same organizations and clubs and seem to agree on most community problems and programs, they may comprise a consensual elite. The more evenly distributed the nominations, the more pluralistic the system is likely to be. The pluralistic system demonstrates more differences and even competition among the leaders.

The *decisional* approach rejects the assumption that a person who holds an important position, or who is perceived by others to be influential, is in fact powerful and does control, dominate, or influence what happens in the community.[11] The decisional approach emphasizes who does what on a specific community decision or issue. One may begin by identifying several recent decisions and who—individuals, groups, or organizations—initiated the issues, what different sets of solutions were proposed, who supported or opposed them, what the bargaining process involved, and who won or lost, or what compromise was developed. One might conclude that if there are just a few key persons initiating issues and establishing proposals that prevail, the leadership is elite. But if many people participate and decisions require compromises among them, then the leadership pattern is pluralistic.

While most leadership studies of small towns have found elite patterns, a number of studies using the decisional approach have found pluralistic patterns. This divergence of findings among those who study community power has led to major quarrels among social scientists over their concepts of power, their identification process of community leaders, the places where decisions are made, and what kinds of community policies are being decided upon. Perhaps no one approach should be relied on exclusively but one method should be used to check out the results of the other. It is for this reason that Swanson has proposed the "multi-faceted" approach which combines the above approaches and presents a far more conditional portrait of the exercise of power and influence. In addition he suggests that it is important to consider the leadership's values, in that "influentials in local communities were found to be leaders of ideological and class interests and their risks and rewards are largely determined by the values of the prevailing group or groups."[12]

Patterns of leadership and influence are affected by the size of the community. Gilbert, in her secondary examination of hundreds of community studies, found population size to be important in the following ways:[13]

1. The larger the size, the more likely the power structure is pluralistic; and obversely, the smaller, the more likely concentrated.
2. The larger the size, the greater is the participation of politicians in carrying out community political functions, and obversely, the smaller, the more likely informals.
3. The larger the city, the less controlled the community conflict.
4. The larger the size, the greater the number of events that become "issues," the greater the number of public-policy decisions, the greater the role of technical expertise as one moves from one decision or scope to another.
5. Rapid growth leads to controversial issues and subsequent withdrawal of economic dominants from participation in community decision-making.

Leadership patterns are also related to population composition and the mobilization of resources. Aiken found that cities having decentralized or dispersed power configurations (pluralistic leadership patterns) are "likely to be more heterogeneous—with more citizens of foreign descent and more children in private schools (an indicator of Catholic influence)—and likely to have a larger working class than cities with concentrated power structures."[14] Furthermore, cities with decentralized power structures were "more successful in mobilizing local resources in order to participate in several federal self-help programs—low rent housing, urban renewal, the war on poverty, and model cities."[15] He notes that each of these decisions is based on acquiring external resources and *not* on reallocations of local community resources.

Patterns of leadership have very specific effects on political processes in a community. Swanson, in contrasting big complex and small simple communities, speculates on the consequences of greater political consensus in small towns as leading to:[16]

> A. fewer participants with one big inner governing circle
> B. stronger incentives to conform to accepted political codes of behavior
> C. higher cost for political dissent
> D. fewer contestants for office
> E. group conflict more likely to polarize the whole community
> F. policy-makers rely for information on:
>   1. colleagues
>   2. friends and acquaintances
>   3. community influentials
>   4. work associates
>   5. those who attend meetings
> G. civic communication patterns are face-to-face and interactive.

Community Resources and Their Mobilization

To have influence in community affairs, one is expected to be a leader. Leaders, however, need access to others in the community. It is important for a leader to have accumulated certain community resources and to be ready and willing to use them to achieve his or her objectives. The following inventory represents the resources which have been found to be commonly used in various combinations to influence community affairs:[17]

> (1) money and credit (employers who set wages and bankers who make loans);
> (2) control over jobs (employers who provide jobs and unions who engage in collective bargaining);
> (3) control over information and mass media (publishers who print the news and reporters who write it);
> (4) high social status and access to leaders (those who are members of a leading family and those who belong to exclusive clubs);
> (5) knowledge and specialized technical skills (lawyers, teachers, engineers, planners, and so on);
> (6) popularity and esteemed personal qualities (athletic heroes, entertainers, and so on);

(7) public office and the legitimacy accorded by citizens (those elected and their appointees);
(8) subsystem groups providing solidarity (religious, fraternal, and ethnic organizations);
(9) the right to vote and participate (to cast a ballot, to appear at public hearings and peacefully assemble to petition);
(10) leadership skills (the strategic and tactical ways to mobilize public support);
(11) manpower and control of organizations (to marshall support from various community organizations);
(12) control over the interpretation of values (leaders who guide the community's decision-making agenda).

These community resources are generally unevenly distributed (a well-accepted proposition) and used with varying degrees of skill (a less-understood phenomenon). Most community activists concern themselves with the accumulation and spending of their scarce community resources in order to achieve their preferred goals and objectives. They give less attention to how they use their access to community resources. Some participants squander their resources while others let it be known they are willing to use their resources for highly selective purposes. Still others hoard their resources, seldom using them but when they do, they do so ineffectively. Therefore the test of the effectiveness of potential influence and the use of that potential in community affairs is the results. If a massive resource base such as great wealth has been accumulated, then simple possession may be enough, especially if having wealth is accepted by others as evidence of power.

In a one-company town the owner actually uses his resources to acquire more resources, and his control over employment limits competition from other influence seekers and inhibits workers from expressing their complaints. A ruling clique in such a community may seldom have to use its resources directly to convince opponents the risk is too great to oppose this clique. However, if resources have been accumulated by various competitive or differing interests, then a wise and selective use of those resources will be necessary in order to compete successfully. If individuals with grievances have little or no apparent resources, they may discover that they can exert some influence by disrupting the community through breaking its laws. They may protest or otherwise embarrass those who govern. These individuals do so, of course, at considerable risk to themselves and their families, especially in small towns. The confrontation strategy has been effective, in some places, in gaining official response to very deeply felt problems.

To bring about community action in small towns generally involves both citizens and leaders. Those who take the initiative and attempt to influence others to participate and follow their lead are generally referred to as mobilizers. Very often they mobilize people through skillful use of the resources within their command. Mobilizers may be elected public officials, civic leaders, ad hoc groups who express the prevailing public opinion, or an external agency or group that wishes to induce change from outside the community. Mobilizers may be self-interested, work for the advantage of a specific group, or have the general community interest at heart. The effectiveness of mobilizers depends on their ability to persuade their followers, coopt or attract those in the middle or on the fence into their camp, bargain with their opposition and even force them by some means to accept the mobilizers' direction.

Mobilization is essentially a process involving the selective marshalling of various resources to influence those in positions of authority (elected public officials); those in community organizations (civic leaders); and interested citizens (attentive public) who all play a major role in forming expressed public opinion. Mobilizers consider how and when to put their issues and preferences upon the community agenda. Care should therefore be given to the stages in the decision-making process. Clark has developed a six-stage model to describe the process:[18]

1) issue recognition—the development of problem statements and the pressure to have something done;
2) information collection—the gathering of facts and opinions about the stated problem;
3) formulation of policies—proposing a solution to the problem;
4) evaluation of policies—exploring alternative solutions and weighing their relative merits;
5) policy selection—determining the most feasible and acceptable solution;
6) policy enactment—translating a policy into operational terms for implementation.

Some mobilizers may be more effective in one stage of the decision-making process than in another. That is, some persons, members of a neighborhood group, for example, are more able to articulate their problems and gain attention (issue recognition), while others, such as planners, can effectively collect, assemble, and display information. Some participants are quick to propose a solution, only to find civic leaders are more likely to play a more prominent role in evaluating the alternative policies. Elected officials must consider the acceptability and feasibility of a particular decision when making the final selection, especially if it involves the use of public monies. Appointed administrators are expected to implement the agreed-upon policy and enforce it at the lowest possible cost and to the greatest satisfaction of the citizenry or, at least, of the influentials.

The budget-making process contains a good set of decisions by which to examine how the various aspects of political influence developed in this chapter actually operate at the community level. Most of the attention of social scientists has focused on the relationship between those in formal office and those with informal power and influence. While the municipal budget is decided upon by elected public authorities, such as the mayor and city council members, those informal leaders who have status and influence often play significant roles in determining the level and priority of public expenditure.

Finally, it is important to understand the relative worth placed on available community resources. While this in part depends on the culture of the town and on who controls what resources, it also should be noted that the value of any one resource may change according to its availability, its general usage, its convertibility to power, and its expendability.

Public Functions and Finance

The functions provided by local government vary considerably among communities. In small towns there is generally a limited scope of government with few public goods and services being provided. Whether the private or the public sector performs the function depends on whether the local government has the necessary funds, or whether it believes itself capable of providing the services, or whether it sees the expenditures as wasteful. Small towners may not be aware of what proportion of activities are actually undertaken by county, regional, state, and federal agencies. That is, although it may appear in any particular community that local government provides few goods and services, closer examination reveals that external governments have assumed a wide range of activities. Table 5.2 lists community goods and services and who provides them or where one goes to receive them.

The table demonstrates how pervasive the role of the public sector can be in our daily lives if the community has chosen to assume responsibility for the many functions listed. It does not, however, evaluate the adequacy of goods and services provided. Despite complaints against the growth of government, those who live in small towns would be without a number of important services if they had to rely only on local government.[19] Furthermore, it should be quickly pointed out that local governments rely heavily on financial aid from state and federal governments.

In recent years there has been an increased interest in transferring the responsibility for performing certain public functions from municipalities, towns, and villages to the county governments, regional districts, Council of Governments (COG), or the state. Zimmerman indicates that "there is growing recognition that certain services can best be provided by a jurisdiction with a larger areal base than a municipality."[20] The most common functions to be transferred during the period 1965 to 1975 were in the order of frequency: solid waste collection and disposal, law enforcement, public health, sewage collection and treatment, taxation and assessment, social services, building and safety inspections, and planning. While small towns were less likely to have transferred functions, about 10 percent of those 2,500 to 10,000 in size plan to do so within the next few years. For those that did, the reasons given were to achieve economy of scale, eliminate duplication, and fill in for the lack of facilities and equipment.

An examination of a municipal budget, which is public information, readily indicates what functions are being performed by local governments and how much money is being spent. It shows the sources of revenues (from where the money comes) and the expenditure pattern (on what the money is spent). The demand for new and additional services is an important aspect of local decision-making because it calls for a shift in the priorities for

Table 5.2:

Goods & Services	Provider			Located In	
	Public	Private	Volunteer	Local	External
Police Protection	x			x	x
Fire Protection			x	x	
Telephone		x			x
Foodstuff		x		x	x
Shelter (housing)	x	x		x	
Clothing		x		x	x
Medical Care		x		x	x
Insurance		x			x
Education	x	x		x	x
Training Program	x	x		x	x
Water	x	x		x	x
Electricity	x	x		x	x
Entertainment		x		x	x
Restaurants		x		x	x
Sewers	x			x	x
Streets	x			x	x
Street Lights	x	x		x	x
Sidewalks	x			x	
Mental Health Clinics	x	x			x
Religion		x		x	
Pest Control		x			x
T.V. Cable		x			x
Appliance Repair		x			x
Auto Repair		x		x	x
Land Use Controls	x			x	x
Certification of Leadership	x			x	x
Gambling		x		x	x
Garbage Collection	x	x		x	x
Recreation/Athletics			x	x	
Parks	x			x	x
Cultural Activities			x	x	x
Pollution Control	x				x
Public Health	x				x
Hospitals/Clinics		x	x	x	x
Laundry		x		x	
Unemployment Comp.	x				x
Postal Service	x				x
Landscaping		x		x	x
Airport	x	x		x	x
Mortician/Cemetery	x	x		x	
Transportation/Bus		x		x	x
Licenses	x			x	x
Courts	x			x	x
Financial Loans		x		x	x
Banking		x		x	x

Table 5.2 (Continued)

Goods & Services	Provider			Located In	
	Public	Private	Volunteer	Local	External
Newspaper		X		X	X
Periodicals		X			X
Library	X			X	X
Household Furnishing		X		X	X
Baby Sitting		X		X	
Flood Control	X				X
Family Counseling			X	X	X
Snow Removal	X			X	X
Ambulance			X	X	X
Pharmaceuticals		X		X	X

NOTE: Prepared in cooperation with Ms. Amy Corton.

municipal revenues and expenditures. This fact should be taken into account by those engaged in community renewal.

To provide the public services deemed necessary, local leaders and voters must decide how to raise the money to pay for them. Generally, about three-quarters of all revenues come from local sources, the other quarter comes from the state and federal government. The bulk of locally derived money comes from the tax on real property.

Almost all property in town is assessed and placed on the tax rolls at some ratio or proportion of the "market," or "just" or "true" or "fair" value. Assessments are made according to a uniform formula by the local or county tax assessor, but the ratio of assessments to market value varies from some 25 to 100 percent depending on the state and/or community. Thus, if a family purchased a $20,000 home, it might have an assessed value of $5,000 if the ratio is 25 percent, or $20,000 if the ratio is 100 percent. Any final assessment may be appealed to a special assessment board. Commercial and industrial property is generally assessed at a higher ratio than residential property. Some properties owned by religious institutions, schools, other nonprofit organizations, or government are exempt from tax.

The actual amount of taxes to be paid by any one property owner is determined by the tax rate expressed in mills per dollar of assessed value, or dollars per hundred or thousand dollars of assessed value. For example, if the tax rate is 10 mills and a family owned a $15,000 home assessed at $5,000, then its tax bill would be $50. Of course, most property is taxed not only for municipal purposes but also for county, schools, and other special districts. Thus, a family may be taxed 10 mills for city tax, 8 mills for county tax, 15 mills for school tax, and 2 mills for a special sanitation district. Therefore, the tax bill would be $50 for the city, $40 for the county, and $75 for the schools and $10 for sanitation, or a total of $175.

The tax rate itself is determined by how much money local governments must raise through property taxation. When making up its budget, public officials—municipal, county, and school—decide the taxes to be raised in relationship to the amount of revenue available from other sources and the amount they plan to spend. Other sources of revenues include state money from taxes on gasoline, cigarettes, alcoholic beverages, and so on, as well as state aid based on formulas for education. The federal government provides outright general revenue-sharing money as well as special grants of money for sewer construction, police equipment, urban renewal, antipoverty programs, and the like. These federal funds are generally made available to local governments upon application. Other general sources of revenue are fees for licenses to conduct businesses, building permits, and fines for violating local ordinances. Another method of securing money is borrowing for the purpose of constructing streets, a city hall, a sewer system, and the like. The borrowed money is paid back over a number of years as debt service which includes not only a part of the money borrowed but also interest charges.

The pattern of expenditures varies considerably from one community to another. However, public expenditures in small towns are generally made in the following order of priorities: (1) highways (maintenance), (2) police protection, (3) sewerage, (4) sanitation (garbage collection), (5) hospitals, (6) education, (7) fire protection, (8) parks and recreation, (9) libraries, (10) housing and urban renewal, and (11) health. Nationwide, interest payments on debt generally equals as much as that spent on fire protection. Shaping the particular expenditure pattern often leads to community controversy. First there is disagreement over the total amount as it affects the amount of local taxes to be raised, and second, over which public functions should be increased or decreased. Most controversy occurs when one group in town wants a new

Figure 5.3: Expected Gains and Losses for Two Community Development Projects[21]

Gains/losses	Public Housing	Downtown Redevelopment
Direct gains	builder suppliers banks tenants	downtown business financial institutions city government builders
Indirect gains	neighborhood merchants	consumers tourists
Negligible gains	middle class	lower class
Net loss	former landlords developers of private housing	dispossessed tenants neighborhood shopping centers

function (hospital, sewer, and so on) established and another opposes it.

In part the controversy may stem from the expected gains and losses from community development projects. Swanson and Swanson speculate on the gains and losses from building a public housing project compared to a downtown redevelopment project. Figure 5.3 identifies a number of trade-offs that may occur in local politics.

Conclusion

There is little agreement about small town politics. Vidich and Bensman note "the pervasiveness of politics in rural life, the unanimity of decision-making, the minimization of decision-making, and the surrender of jurisdictional prerogatives."[22] Presthus is disturbed "that despite high levels of popular education, economic stability, a fair degree of social mobility, a marvelously efficient communication system, and related advantages usually assumed to provide sufficient conditions for democratic pluralism, the vast majority of citizens remain apathetic, uninterested, and inactive in political affairs at the community level."[23] Wildavsky, in his study of a university town, is more hopeful as leaders take into account the preferences of the activists and are influenced by their expectations of what the community will accept.[24]

Perhaps the best summary on the study of community influence and decision-making in small towns would be to list the propositions developed by Presthus in his detailed examination of two small towns in upstate New York.[25] He found support for the following statements:

1. There is an inverse relationship between overlapping membership (elitism) and size.
2. In communities with limited leadership and economic resources, the power structure will more likely be dominated by political leaders, whereas in those with more fulsome internal resources, it will probably be dominated by economic leaders.
3. There is a positive relation between the degree to which a community is socially integrated and the manner in which it solves its problems, i.e., through some citizen participation in crucial local decisions or through more centralized control and action by a few hyperactive leaders.
4. Community institutions (hospitals in this case) which enjoy the closest ties with the members of the power structure received greater support and operated at a higher level of competence.
5. The better organized a social structure (class, ethnic group, residential area) is under its own leadership, the more politically effective it will be.
6. Among "non-political" associations, upper middle-class associations are more likely to focus attention on civic and political affairs than those of any other class.

Community analysis then can assess the nature of the local political system in terms of decision-making arrangements, patterns of influence, the mobilization of community resources, the prevailing scope of local government and some of the key aspects of public finance. Any community renewal project should include understanding the local political system as it generally requires not only a decision of local government or key community organization but sufficient support to implement project objectives. Therefore it is most important that those who propose and implement community renewal activities know the key influentials, their way of doing things, as well as the local possibilities and constraints imposed by local resources.

DISCUSSION GUIDE

I. *What is the rate of local participation in national and local elections?*

   (A) What proportion of the population is eligible to vote (those 18 years or older)? For places 1,000 to 10,000, see by individual state *General Population Characteristics,* Bureau of the Census, 1970.

   (B) What proportion of the eligible people are registered to vote? Consult County Board of Elections or Town Clerk for number of registered voters in precinct or election districts in town.

   (C) What proportion of the registered people actually voted in the last elections for:
      —president (1976)
      —governor
      —mayor
      —town council

It is commonly believed that political participation is inversely related to size. Yet, two quite contradictory theories have been proposed. One is the mobilization theory that predicts greater participation in urban settings where there is exposure to more communications, more interaction, and more encouragement from others to become involved in politics. Milbreath, for example, finds that "persons near the center of the society are more likely to participate in politics than persons near the periphery.... [P]ersons near the center receive more stimuli enticing them to participate, and they receive more support from their peers when they do participate."[26] The alternative theory predicts a decline in participation as one moves from the smallness and intimacy of small towns to the massive impersonalization of the city. Verba and Nie state, "In the small town, the community is a manageable state.

Citizens can know the ropes of politics, know whom to contact, know each other so that they can form political groups. In the larger units, politics is more complicated, impersonal, and distant."[27] They also find that not only does participation decline as communities grow in size, but even more so as "they begin to lose the clear boundaries that separate them from other communities. Participation in general and communal participation in particular are more widespread in more peripheral and isolated places. As one moves to the 'center' of society, such activity is inhibited."[28] As a result, they believe it important to distinguish between the following types of communities: (1) isolated villages and rural areas, (2) isolated towns, (3) isolated cities, (4) small suburbs, (5) large suburbs, adjacent cities, and (6) core cities.

II. How well attended are public meetings and functions?

A. Regular town council meetings? _____
B. Public hearings on special decisions, on budgets, projects, and so on? _____
C. Ceremonial events such as the Fourth of July, parades, and so on? _____

(Ask the town clerk to see the minutes of meetings or to estimate attendance at each type of meeting or consult the local newspapers or someone who regularly attends. Use actual numbers or terms as large, medium, and small attendance.)

If a community has a high degree of citizen participation as measured by voting and attendance at public meetings, it is more likely to be pluralistic and open. The more people participating in public affairs supposedly reflects a variety of expressed points of view at the decision-making centers which stimulates public officials to respond. Lower rates of citizen participation are not only associated with elite leadership patterns but often indicate that there is a selective bias on behalf of the more affluent, well-educated, higher status people who are generally more involved than the poor, unemployed, young, and disadvantaged residents. However, one must watch for the situation where participation is habitual and where almost all the key decisions are made *before* the vote or other means of participation has taken place.

III. What is the form of a local government and the method of electing public officials?

Consult local public officials. This information is also listed for many small towns in *The Municipal Year Book 1968* (The International City Manager's Association, 1968), pp. 52-130.

A. What classification is local government?   Community
   —municipality
   —town
   —township
   —borough

B. Form of government?
   1. weak-mayor
   2. strong-mayor
   3. council-manager
   4. town meeting
   5. representative town meeting
   6. commission

C. Do candidates run on a ballot that is:
   (1) partisan
   (2) nonpartisan

D. What is the term of the mayor? Number of years

E. What is size of the city/town council?
   Number of members

F. How are the members of the city/town council elected?

   at large
   by wards or districts
   both

G. What is term of the city/town council?
   Number of years

The state legislature grants certain authority to local governments according to their classifications. Municipalities generally are granted more authority to perform more functions. Most towns (61.7 percent) with populations 5,000 to 10,000 use the mayor-council form of government and the remaining one-third (34.0 percent) use the council-manager form. Most small towns also use the partisan ballot with national political party labels identifying candidates. Most council members are elected by at-large constituencies.

IV. What is the prevailing leadership and/or influence pattern?

It is sometimes difficult to know who governs a community. Some people would simply say the elected local officials; others believe economic elites dominate public officials and the community in general. We suggest using one or more of the following approaches discussed in this chapter.

A. *Positional.* Compile a list of persons who are on the governing boards of community decision-making bodies and organizations.

Town/City Government	Chamber of Commerce

School Board	Other

Some considerations would include whether there are many or few persons in visible positions of community decision-making, and whether some few individuals hold more than one.

B. *Reputational.* Ask a few very knowledgeable people whom they believe to be people who have influence in community affairs.

1. Most generally influential in town.

   _____   _____   _____
   _____   _____   _____

2. Best person to go to for advice about a community project.

   _____   _____   _____
   _____   _____   _____

3. Most likely to support a specific proposed project and help implement it through use of their influence.

_____   _____   _____
_____   _____   _____

Rank order the people by the number of times they have been mentioned. If only a very few people receive most all the nominations the leadership pattern is likely to be elite. Notice if some people will be mentioned as having only a *specific* role while others are generally influential.

C. *Decisional.* One can take a recent community decision and ask the following questions.

1. Who initiated the proposed action?

_____   _____   _____

2. Who supported and/or opposed the proposed action?

Supporters	Opponents
_____	_____
_____	_____

3. Who negotiated a resolution so the decision was made?

_____   _____   _____
_____   _____   _____

4. Whose views tended to prevail?

_____   _____   _____

The fewer people or groups mentioned in each of these stages of decision-making, the more likely an elite system exists.

In trying to determine from the information gathered what kind of leadership pattern exists one should count how many leaders there seem to be, how well-known they are to the general public, what scope of influence they seem to have, how much issue agreement there seems to be, and how much continuity there seems to be in the prevailing leadership pattern. Figure 5.4 illustrates the tendency of these characteristics if placed on a continuum to be more associated with an elite or pluralistic leadership pattern.

Figure 5.4: Characteristics of Elite and Pluralistic Leadership Patterns

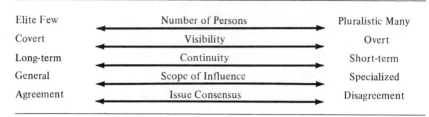

There are a number of factors that are often associated with the two predominant patterns of influence. For example, a small, independent, stable community that has a homogeneous population, few community organizations, a selective inadequate resource base, and has a dominant company (industrial firm) that uses nonunion labor would more than likely have an elite pattern. Similarly, the town would more than likely have a city-manager form of government providing minimal public services. Elections would be at-large and nonpartisan (or one party control). Low levels of citizen participation and little conflict over community issues would exist. Perhaps most important but more difficult to determine is the role of the informal leaders, who do not hold public office. They are generally vaguely known to the general public and form a small inner clique which holds general influence in almost all communitywide policy-making.

It should also be noted that there is a tendency to pass judgment on which type of pattern of influence is best. The answer depends on what people believe is in their best interest. Elite systems have a reputation for being *effective* in making decisions, on the one hand, but being removed from the expressed will of the majority on the other. The pluralistic systems have a reputation of being more open and *responsive* to citizen preferences, but they are also less likely to make quick decisions, sometimes to the point of being paralyzed or unable to take action.

V. What community resources are held or controlled by whom?

	Name	Name	Name
Money & Credit	1._____	2._____	3._____
Jobs	1._____	2._____	3._____
Information/Mass Media	1._____	2._____	3._____

Status and Access to Leaders	1._____	2._____	3._____
Knowledge & Specialized Skills	1._____	2._____	3._____
Popularity & Personal Qualities	1._____	2._____	3._____
Public Office	1._____	2._____	3._____
Social Group Support	1._____	2._____	3._____
Voting and Participation	1._____	2._____	3._____
Leadership Skills	1._____	2._____	3._____
Organizational Resources	1._____	2._____	3._____
Interpretation of Values	1._____	2._____	3._____

Particular community resources are very important in the small town. They involve socioeconomic and political resources that are constantly in the forefront of small town issues. One important resource is the concentration of wealth, which in small towns is most visible in the ownership and use of land and buildings. By referring to the municipal or county tax records, which lists properties in town by location, owner's name, and amount of taxes paid, one can determine who are the town's biggest taxpayers and how many residential and commercial properties they own. The landowners' sensitivity to changes in the tax rate gives them a "stake" in municipal governance. Another very visible aspect of economic resources in the context of community power and influence is the control of local commerce and industry. Although information is collected at county and regional centers, the size of the small town lends itself to "in the field" observation. A resident could visit each local business and ask the proprietor or manager for estimates of gross sales and number of employees. The number of employees provides a businessman with some influence over a number of families. This influence may, in many cases, be a more significant resource than the total value of the firm's production.

Patterns of Influence and Decision-Making in Small Towns

One should also evaluate how community resources are used. One participant may wield greater influence than another, or a small businessman may be more skillful in exercising influence than the manager of a large firm because he has political "savvy" and knowledge of how to use economic resources effectively for political purposes. On occasion, the influence of large landowners can be overshadowed by the influence of a small landowner whose properties are located in pivotal spots (such as downtown) and are used to the utmost political gain.

VI. What are the major sources of revenues for municipal or town government?

Secure a copy of the local municipal or town budget from an official at city hall. Identify how much money is received from the state and federal governments (see Table 5.3). Then determine what amounts come from local sources such as property taxes, other taxes, fees, fines, and so on. Note to what extent the community is self-reliant and to what extent intergovernmental transfers are used. It should be noted that state law specifies that

Table 5.3: Per Capita General Revenue Patterns for Small Municipalities and Townships—1972

	Cities by Size			Township by Size*			Community**
	5,000–9,999	2,500–4,999	Less than 2,500	5,000–9,999	2,500–4,999	Less than 2,500	
General Revenue	120.14	106.38	91.40	133.14	97.74	88.33	_____
Intergovernmental aid	27.71	25.17	20.89	32.46	26.14	27.32	_____
General Revenues Own Source	92.41	81.28	70.51	100.68	71.60	61.01	_____
Taxes	57.71	49.44	40.23	90.13	65.66	55.05	_____
Property	39.39	33.83	30.72	83.79	59.17	50.41	_____
Charges, etc.	34.68	31.85	20.28	10.55	5.94	5.97	_____
Debt Outstanding	226.81	231.49	154.05	111.99	52.13	56.57	_____

*Townships in the 11 strong township states
**Per Capita = Dollar expenditures for each item ÷ Total Population

SOURCE: U.S. Bureau of the Census, *Census of Government*, Volume 4, Finances No. 4; *Finances of Municipalities and Township Government* (Government Printing Office, 1974).

certain amounts of gasoline taxes must be spent to pave and maintain local streets. Also note what amounts are received from the federal government for specific projects and "general revenue-sharing" funds. Analysts of public finance have come to believe that heavy reliance on relatively inflexible local property taxes places cities in the difficult position of not raising sufficient money to meet the increasing costs of employees' salaries, materials, and services. They suggest that more use should be made of charges for specific services received, such as parking meters, licenses, and so on. Local public officials are very reluctant to raise local property taxes because citizens often express their disapproval face-to-face with the officials at public hearings and at the election polls.

VII. What is the expenditure pattern of municipal government?

Calculate from the budget what proportion of the total budget goes for each function (see Table 5.4).

Table 5.4: Pattern of Local Per Capita Public Expenditures for Small Municipalities and Townships—1972

	Cities by Size			Township by Size*			Community**
	5,000–9,999	2,500–4,999	Less than 2,500	5,000–9,999	2,500–4,999	Less than 2,500	
General Expenditure	123.69	114.10	91.92	130.98	96.19	85.88	_____
Highways	19.88	20.49	19.67	21.57	23.94	35.97	_____
Hospitals	6.91	3.24	.04	NA	NA	NA	_____
Health	.52	.39	.08	NA	NA	NA	_____
Fire	6.81	5.30	1.28	4.06	3.47	.67	_____
Police	17.75	16.34	12.20	7.51	4.53	2.27	_____
Sewerage	15.36	14.64	2.24	8.16	3.59	1.02	_____
Sanitation	7.22	6.38	1.11	2.57	1.64	.29	_____
Parks and Recreation	5.79	4.70	.87	2.65	1.51	.30	_____
Housing & Urban Renewal	1.89	1.14	–	NA	NA	NA	_____
Libraries	1.65	1.22	.25	NA	NA	NA	_____
Interest on Debt	6.06	6.58	6.27	3.82	1.81	1.10	_____
Other	13.26	18.75	35.23	11.69	8.64	21.33	_____

*Townships in the 11 strong township states.
NA = not applicable
**Per Capita = Dollar expenditures for each item ÷ Total Population

SOURCE: U.S. Bureau of the Census, *Census of Government*, Vol. 4, No. 4; *Finances of Municipalities and Township Government* (Government Printing Office, 1974).

These expenditures reveal local priorities and reflect the values and interests of those who participate in community affairs either directly in elections or indirectly through influence with those in authority or with civic leaders. Intense struggles sometimes occur over how the town should spend its public money. This is especially the case when a new program, project, or activity is proposed. In some communities a very small elite determines the budget. In other places the final outcome is influenced through keen and vocal concerns aimed at public officials.

NOTES

1. Max Weber, *The Theory of Social and Economic Organization* (Oxford: Oxford University Press, 1947), p. 152.
2. George A. Hillery, *Communal Organizations* (Chicago: University of Chicago Press, 1968), pp. 185-186.
3. For a systematic listing of associations operating in American Communities see Roland L. Warren, *Studying Your Community* (New York: Russell Sage Foundation, 1955), pp. 287-288.
4. Bernard Berelson and Gary A. Steiner, *Human Behavior* (New York: Harcourt Brace & Jovanovich, 1964), p. 380.
5. Adapted from Sidney Verba and Norman H. Nie, *Participation in America* (New York: Harper & Row, 1972), p. 31.
6. Ibid., p. 97.
7. From a study of the 1968 and 1972 presidential elections, see U.S. Bureau of the Census, *Current Population Reports* Series P-20, Nos. 192 and 228.
8. For a four-fold set of diagrams of "monoliths," "polylithic," "mass participation," and "pluralistic," see Terry N. Clark, *Community Structure and Decision-Making* (San Francisco: Chandler, 1968), pp. 38-43.
9. A specific example of the positional approach is Robert O. Schulze, "The Bifurcation of Power in a Satellite City," in *Community Political Systems,* edited by Morris Janowitz (New York: Free Press, 1961), pp. 18-80.
10. While there have been reputational studies, one of the more careful studies of small towns was conducted by Robert Presthus, *Men at the Top* (Oxford: Oxford University Press, 1964).
11. For example, see Aaron Wildavsky, *Leadership in a Small Town* (Totowa, NJ: Bedminster Press, 1964).
12. See Bert E. Swanson, "Community Power: Concepts, Methods and Interventions," in Donald Freeman, ed., *Foundations of Political Science* (New York: Free Press, 1977), p. 367.
13. C. W. Gilbert, *Community Power Structure* (Gainesville: University of Florida Press, 1972), pp. 38-46.
14. Michael Aiken, "The Distribution of Community Power," Michael Aiken and Paul Mott, *The Structure of Community Power* (New York: Random House, 1970), p. 514.
15. Ibid., p. 515.
16. Bert E. Swanson, "Small Town and Big City Politics," in *Towards Smallness: A*

Human Perspective for Human Scale, edited by Harold S. Williams (Rodale Press, forthcoming).

17. See Robert Dahl, "The Analysis of Influence in Local Communities," in *Social Science and Community Action,* edited by Charles Adrian (Institute for Community Development and Services, 1969), p. 32. See also Terry N. Clark, 1968, pp. 38-43.

18. Clark, op. cit., p. 65; for a process that accounts for the decisions of total government, see Agger, Goldrich and Swanson, *The Rulers and The Ruled* (Duxbury Press, 1972), pp. 24-28; (1) policy formulation; (2) policy deliberation; (3) organization of political support; (4) authoritative consideration; *Event*: decisional outcome; (5) promulgation of the decisional outcome; and (6) policy effectuation.

19. For a survey of local public officials on who should provide many of these services, see Joseph F. Zimmerman, "Intergovernmental Service Agreements and Transfer of Function," *Substate Regionalism and the Federal System Vol. III* (Advisory Commission on Intergovernmental Relations, 1974), pp. 183-184.

20. Joseph F. Zimmerman, *Municipal Transfers of Functional Responsibilities,* Urban Data Service Report (September 1975), p. 1.

21. Bert E. Swanson and Edith Swanson, *Discovering the Community* (Irvington Publishers, 1977), p. 341.

22. Arthur J. Vidich and Joseph Bensman, *Small Town in Mass Society* (Princeton University Press, 1968), p. 109.

23. Presthus, 1964, pp. 432-433.

24. Wildavsky, 1964, p. 360.

25. Presthus, 1964, pp. 408-416.

26. Lester Milbreath, *Political Participation* (Chicago: Rand McNally, 1965), pp. 113-114.

27. Verba and Nie, *Participation,* p. 231.

28. Ibid., p. 242.

Chapter 6

PROFILING COALSTREAM: COMMUNITY

ANALYSIS OF A SMALL TOWN

The example of Coalstream, a small town of some 1,500 people in a state in the Northeast region of the United States, highlights the elements and dynamics of our community analysis framework for small communities. By no means is Coalstream (a fictitious name) to be considered a typical small town. The circumstances, problems, and potentials faced by residents of Coalstream are somewhat site-specific. But the methods of investigation and analysis of this town can be applied in other small communities.

Coalstream has been selected as the focus of our attempt to apply the concept of community analysis to a particular small town inasmuch as we participated in the efforts to renew the community. The data and discussion reported herein centers on what we learned in the early phase (1974) of a community development project which is referred to herein as the Coalstream Project.[1] Some of the materials come from readily available census and other governmental reports. These have been supplemented with interviews, attitudinal surveys, and observations. To draw this community profile we have followed the discussion guide developed in Chapters 2 through 5 in a somewhat different order. That is, we begin with the economy, then proceed to the social and political aspects, and end with the value component. Coalstream residents might well differ with our interpretation, since much has happened in town in the last five years as a result of the Coalstream Project.

The Economy

The starting point for many conventional community renewal programs is the economy. Boosterism and bootstrapping are hallmarks of small town revitalization plans. Designing downtown streets scapes, altering municipal tax structures to attract new businesses, clearing vacant lots and proclaiming them "industrial parks" comprise the immediate responses of many towns to the need for changes in the local economy. Chapter 4 describes the wishful thinking and tough realities of small town economics.

The dream of economic redevelopment hung over Coalstream. All segments of the population were concerned in one way or another. Shop owners on Main Stret railed against inadequate parking for potential patrons. Entrepreneurially minded people talked about opening new businesses in vacant storefronts. For some, an awareness of industrial development elsewhere in the region held both promise and frustration, an image of what Coalstream really might be able to do and what it might never achieve. Old-timers spoke vaguely of "the syndicate," the forces outside the community, located in big cities, which kept the community down. An examination of several indicators of Coalstream's economy confirmed the suspicions of some, confounded others, and provided new grist for the town's economic debate.

Indicators of Employment and Sources of Income

In 1970, only 43 percent of Coalstream's adult population, fourteen years old and older, were in the labor force. In other words, less than half of the community's adult population were gainfully employed. Fully 44 percent were neither employed, unemployed, in school, nor in the service; that is, they were out of the labor market entirely, either never having entered the market (e.g., housewives) or retired from it. For every employed woman, there were three employed males. The role of women in the Coalstream economy was not very significant in direct terms (though, indirectly, the work women do as housewives enables many men to work outside the home).

Coalstream was predominatly a blue-collar town. Almost 71 percent of the employed labor force worked as craftsmen, laborers, and equipment (transportation) operators. They were employed primarily in manufacturing (see Table 6.1). A comparison of Coalstream to the nation reveals sharp disparities: nearly twice the national average of workers are employed in the manufacturing sector.

The economic condition of Coalstream is further clarified in Table 6.2 which provides a breakdown of the population dependent upon income transfers from governmental sources. These statistics omit some other categories of dependent households. Another 21 households received what was called "general relief." The title of that income source caused Coalstream residents to refer to all people receiving public assistance as "reliefers." Also, not included in the statistics are those households receiving railroad or miners' pension benefits or Black Lung assistance. Even taking into account possible duplication (certain families receiving more than one of these public welfare income sources), a full quarter or more of Coalstream's permanent household population relied on government to provide what their own earning powers could not.

Table 6.1: Employment by Industry

	Coalstream (1970) %	U.S. (1970) %
Manufacturing	44.1	24.4
Mining	1.1	0.8
Construction	5.4	5.5
Transportation/Public Utilities	12.3	6.3
Wholesale/retail trade	18.3	18.9
Finance, Insurance, Real Estate	2.6	4.7
Public Administration (government)	1.7	5.2
Professional Services	4.7	16.5
Business/Repair Services	1.1	2.9
Personal Services	3.4	4.3
Agriculture/Forestries	0.0	3.5
Industry not reported	0.0	6.2

Table 6.2: Source of Government Income Transfers

	U.S. (1973) Population %	Coalstream (1973) No. of Households (case counts)	Households %
AFDC	5.3	35	6.8
Social Security	13.5	79	15.4
Old Age Assistance	0.9	16	3.1
APTD	0.6	5	1.0

The actual level of income in Coalstream supports the impression of deprivation and poverty suggested in the analysis of sources of income. The 1970 median family income was $7,187, or $2,680 less than that for the nation. The median income of unrelated individuals in Coalstream was $1,382, or less than half the amount of their counterparts elsewhere in the United States.

To sum up, Coalstream was a lower-income small town with a combination of blue-collar occupations and a relatively high percentage of households dependent on governmental support. The difficulty in earning an average income makes it a lower-status community. Surrounding Coalstream were a few higher-income communities. But overall, the region was little better off than Coalstream itself as the county ranked third out of 67 counties in the state in percentage of the population receiving welfare benefits.

Economic Base

As with many small towns, Coalstream is so small that economic base analysis at the community level does not reveal much. For example, the ratio of Coalstream's basic employment to all other employment is 205 to 260, or almost 1 to 1.27. In other words, for every four basic industry employees in Coalstream, one might expect to see five nonbasic workers. However, that is a relationship without much significance given Coalstream's size. The reason is that most of these workers work out of town at factories along the highway or in the two nearest cities. The presence of basic and nonbasic workers in Coalstream has less to do with its economy than it does with the economy of the county or region. The bulk of Coalstream's industrial work force worked at a plant nearby making glass containers and caps, by far the largest employer in the region with over 2,000 workers. Others worked at smaller plants along a stretch of highway between the county's two largest communities where tank trailers, steel scaffolding, and fabricated pipes were constructed. Still others worked for the water and electric utility companies.

Local employment in Coalstream was negligible. The only two employers of any significance in town were a firecracker assembly plant, employing some dozen women, and a very small machine shop with few employees. The railroad yard at the north end of town was in the process of being phased out, employing only a few men to oversee its limited operations. Other local employment consisted of a few clerks and cooks.

The use of an economic base measurement for Coalstream is inappropriate as can be demonstrated by the example of a longstanding economic development plan in the region. A dozen miles away was a planned site for a large automobile assembly plant. Despite the distance, many Coalstream laborers expected to get jobs at this plant when it opened. Assuming they did, the proportion of Coalstream's basic industry employment in the total community work force would have gone up dramatically. On paper, the economic picture of Coalstream would move toward basic employment, suggesting that through the multiplier effect there would be an increase in the town's nonbasic employment. Realistically, what would happen, of course, is that Coalstream's nonbasic picture, its service jobs and businesses, would not grow at all. For the region as a whole, the new auto assembly plant would undoubtedly generate a significant shift in nonbasic employment. For the community lucky enough to host the plant, the situation would be particularly significant. But for Coalstream, with its work force shifting jobs out, not adding any new ones, especially in town, the multiplier would be negligible.

The happenstance of basic and nonbasic industry workers living in a particular community is of little consequence unless the employer's factory or plant is also located there. Extending the boundaries of analysis to a size including both the residences of most of the workers and the location of the plant makes economic base analysis much more meaningful. At the county level, 17,743 persons were employed in largely "export" industries (manufacturing, mining, agriculture, and forestry), compared to 28,468 in other, more locally oriented, service jobs, or a ratio of 1 to 1.6.

Location Analysis

The first questions to be posed in examining Coalstream's locational advantages are what are and where are its raw materials? For many years, mining was the mainstay of the local economy of Coalstream and several other towns. A very high-grade coal seam stimulated not only active deep mines, but numerous coke ovens and foundries. At approximately, the time of the Great Depression, the best seams were being exhausted and deep mining was being replaced by the less labor-intensive truck and strip mining. As the coal business declined, so did its companion industries, particularly the coke ovens and the railroads. Despite a few spurts of life in the industry over the years, its steady decline saw coal disappear as a major economic force. Aside from a bit of bluestone in the hills behind Coalstream, there was little bounty of nature to act as a drawing card for Coalstream's economic attractiveness.

At first glance, Coalstream's labor force is also not an attraction. Skilled workers and trained professionals are scarce. Overall, the population is structurally outside the normal labor market. But beneath the surface, there are potential strengths. That so many maintain successful work records at out-of-town sites means that they might become just as good a group of employees for a locally situated plant. There is also a manpower potential among those not presently in the labor force. There are in the nonworking population several groups of people with energies and skills to offer. Women generally comprise part of Coalstream's surplus of labor theoretically available for employment in new industries. Similarly, the town's many retirees are in that reserve. Though societal mores and laws may have forced them to retire from their regular jobs at an arbitrary age, their skills and earning potential may be usable in new businesses. That many people concerned about economic development do not fully inventory the skills of the surplus labor force is a serious limitation in the range of strategies they adopt.

Situated at the base of a mountain ridge in an area considered quite remote by urban dwellers, Coalstream has certain assets in its geography. On the north-south axis, it is located sixty miles from a city of nearly one million, and twenty miles from a city of almost one hundred thousand. On an east-west axis, it sits in the middle of the county's two largest cities and major employment centers. The highway between those communities, built as a result of a regional growth center strategy of public infrastructure investments, has successfully lured to the area several major industries, a community college, and two shopping centers. But the road from the highway into Coalstream is a steep, winding route, leading to a dead-end at Coalstream's tiny downtown. Although some heavy trucks occasionally make the trip down the road to get to the railroad yards, the road is not exactly ideally suited to major truck traffic. Also, the road meanders through quiet residential areas. The railroad still has its freight trains pass through town. Passenger stations were closed decades ago. Aside from long coal trains, the railroad does not receive much use. Although there is some coupling and uncoupling in the yards, the main switching place for rail traffic is in a city six miles away.

Even if the highways and rails were in good shape, there are not many sites within Coalsteam where a new factory of any significance could locate. Most of the town's vacant land is at an exceedingly steep grade, at points, a forty percent slope, wholly inappropriate for many industries. Other vacant parcels, usually small, sit surrounded by family residences. The acquiescence of residential areas to industrial development and its concomitant traffic, noise, and pollution is as unlikely in Coalstream, as it is in most communities.

Drawing on nearby tourism is another factor worth considering. Located to the south of Coalstream is a picturesque little community known for its white-water sports and active tourist industry. For Coalstream to take advantage of tourism already developed nearby, the route into the town would have to be improved to draw traffic through Coalstream. The bumpy mountain road is, however, not Coalstream's responsibility but that of the county, and thus an unlikely focus for economic development action.

Local taxes are another factor that might attract business development. Coalstream maintained a very low tax rate—$24.36 per $1,000 of assessed valuation for municipal, county, and consolidated school district together. Assessments were very low and almost never updated. This allowed the town's top landowners to pay as little as $200 to $400 a year in total real estate taxes to the municipality for as many as four or five properties. Although a wage tax was charged to employed residents, employers were taxed only on their personal incomes if they were town residents. If the owner lived outside Coalstream, he or she could escape with paying only the minimal property tax. Such a tax structure is certainly attractive to many investors. However, most other small towns in the county offered similar tax burdens. Coalstream's only additional public inducement to investment comes from the municipality's ownership of the water system. Conceivably, the community could offer special water rates to industries interested in a Coalstream location. Other than these two factors associated with the functions and prerogatives of the municipal government, Coalstream offered little else to lure business. Industrial zoning and designation of development sites achieve little if the potential development is economically unrealistic. With no organized citizen effort to show the benefits of a Coalstream business location, most investors do not give this town much consideration.

Functional Analysis

Coalstream is no longer a coal town. The once heavily mined community now contains only a handful of residents working in the strip operations in the hills, or in the still flourishing coal mines in the next county to the west, some thirty or forty miles away. Coalstream's image is one of remoteness and isolation, closed off from the rest of the world by the hills and mountains around it. But it is not quite as isolated in reality as insiders and outsiders think. Actually, most of Coalstream's work force earns its living outside the town, making it a satellite town but at least giving its residents some exposure. In a town not much larger than a square mile, less than 10 percent of the work force can walk to work. Lacking public transportation beyond a rickety old bus on an irregular schedule, those who work must drive.

The economic reality of Coalstream does not fit the nostalgic image of self-sufficient small towns. The key to its future is locked into the development of the surrounding communities and region. Coalstream might consider improving its vacant land as industrial sites, offering tax incentives in addition to its already low tax structure, and providing some vacant space downtown for office development. But basically the town is a bedroom suburb for commuters to nearby employment centers. Local economic development activities that are not coordinated with regional efforts are likely to be marginal even if successful. Some small scale economic development, perhaps focusing on cottage crafts and other mechanisms for making the economically marginal population of the town productive, are worthwhile activities. But this cannot be viewed as a solution to the town's economic distress and poverty. It should be noted that the failure rate of new small businesses is quite high, making that economic strategy precarious. By itself, the inefficiency of Coalstream's size and the inadequacy of its resource and geographical endowment make economic development strategies limited to the town's boundaries fraught with problems and weak in potential benefits.

Summary

Coalstream has virtually lost its economic base which relied on extractive industries and manufacturing. The economy is now almost entirely dependent on external regionally based firms employing the resident members of the labor force. The town remains unattractive to economic firms who have been asked to locate there. The community does not provide a sufficient consumer's market for a local commercial center as there are already a number of large shopping malls located nearby. Therefore, the town is gradually drifting toward becoming a small residential community with little economic base of its own.

The Social Structure of Coalstream

City dwellers would like to believe the nostalgic "just folks" image presented by small towners to the outside world. It is a hopeful alternative to the turbulence of the city and the motivation for many participants in the recent reverse migration to small towns. The peacefulness of that image does not fit the reality. Even homogeneous small towns, comprised of largely one ethnic and religious group, will develop petty factions around personalities or life-styles. Few self-proclaimed "one happy family" communities maintain that condition over time.

Differences within Coalstream surfaced quite easily, especially around the activities and intentions of the Coalstream Project. Residents held various images of the community dimensions and sources of intracommunity differences. The data generated in

Table 6.3: Annual Incomes of Families and Unrelated Individuals—1970

Income	U.S. Families %	Coalstream Families %	Coalstream Families and Unrelated Individuals %
LT $2000	4.6	4.9	22.1
2 – 2999	4.3	4.7	7.2
3 – 3999	5.1	7.0	5.2
4 – 4999	5.3	13.1	10.7
5 – 5999	5.8	4.1	4.3
6 – 6999	6.0	14.6	12.1
7 – 7999	19.9	23.0	4.6 ⎫ 17.1
8 – 9999			12.5 ⎭
10 – 11999	26.8	22.3	9.3 ⎫ 16.5
12 – 14999			7.2 ⎭
15000 or more	22.3	5.7	4.3

examining Coalstream's social structure substituted factual analysis for subjective impressions.

Socioeconomic Status

Chapter 3 presented three indicators of socioeconomic status—income, education, and occupation. The income distribution in Table 6.3 reveals considerable differences between the income distribution in the nation as a whole and that in Coalstream. It is important to note the following:

(1) Below $4,000, the proportion of Coalstream families was only slightly more than 2 percent greater than the whole country. At the income level of $4,000 to $4,999, the disparity between Coalstream and the rest of the United States was almost 8 percent. At incomes of $4,000 to $6,999 were 17 percent of U.S. families, but a full 32 percent of the families of Coalstream. (The community had 11.8 percent of its families in poverty compared to 10.7 percent in the nation.)

(2) The middle income and wealthier groups, those making more than $15,000 a year, were few in number, representing only 5.7 percent compared to 22.3 percent in the nation.

(3) Ninety-four of Coalstream's one hundred thirty unrelated individuals relied on incomes of less than $2,000, a small amount of money from which to pay rent, buy food, and purchase clothing. No unrelated individuals earned more than $7,000.

Table 6.4: Occupations of Those in the Labor Force—1970

Category	U.S. %	Coalstream %
White Collar,		
Professional, technical	14.8	3.6
Managers, officials, proprietors	11.4	8.3
Clerical and sales	24.7	11.8
Blue Collar		
Craftsmen, foremen	13.5	23.0
Operatives	17.0	30.7
Service workers	10.7	6.4
Farm workers	4.0	0.0
Laborers	4.1	10.5

One might infer from these data that this low-income group is comprised largely of retirees on fixed incomes from social security and railroad pensions. Combined with Coalstream's family-based poor, they tip or "skew" the income distribution toward the low end, forming a distinctive cluster. Only four families earned more than $15,000 annually, and that in households where both husband and wife worked.

A second indicator of socioeconomic status is occupation. Having noted that Coalstream was basically a blue-collar town Table 6.4 contrasts the actual occupations of Coalstream's workers with those of the nation. Occupations relate to status in two ways—by the prestige of the occupation and by the level of income with which they are associated. Coalstream's occupational

Table 6.5: Educational Attainment—1970

Years of School Completed	U.S. %	Coalstream %
Elementary		
0 – 4 years	5.4	5.7
5 – 7 years	9.1	14.6
8 years	13.4	11.7
High School		
1 – 3 years	17.1	15.6
4 years	34.0	45.0
Post-high school or college		
1 – 3 years	10.2	5.0
4 or more	11.0	1.9

distribution is heavily weighted toward the lower prestige jobs. While there were not many professionals in town, the proportion of managers and proprietors closely parallels the average for the whole country, but clerical and sales workers were only half the nation's proportion. The proportion of craftsmen, operatives, and laborers was well above the national average.

Education is the third element of socioeconomic status. Table 6.5 indicates that the educational attainment of adults 25 years and older in Coalstream parallels the national distribution through high school, but few have gone beyond to higher education. The

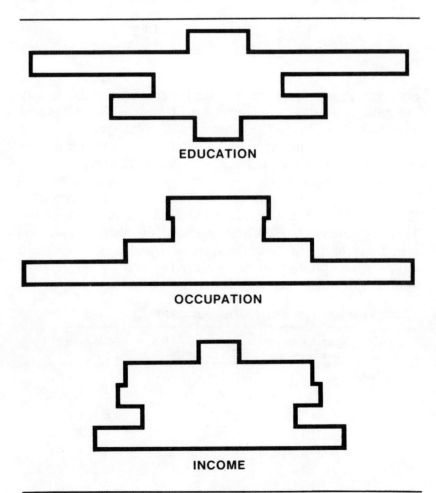

Figure 6.1: Shapes of Coalstream's Social Status Profiles

location of a community college nearby may make a difference in educational attainment for the town's current younger population.

The main feature of Coalstream's SES profiles is that the largest proportion of persons in the lowest strata for income and occupation (see Figure 6.1). Educationally, most residents have attained a higher status as having attended or graduated from high school. Another way to see how Coalstream compares with the national averages is to compare the percentage of Coalstream's residents that fall into each category with the average percentage in the nation. Table 6.6 reveals how much and in what direction Coalstream residents deviate from the national averages. Coalstream falls short of national norms on many parts of each major indicator. The negative percentages in most of the highest or most prestigious categories signify its low socioeconomic status.

Cultural Factors of Differentiation

Race and Ethnicity. With no blacks in the community, race is not a measure of differentiation Coalstream, though negative refer-

Table 6.6: Coalstream's Deviation from National SES Indicators

SES Indicators	Coalstream's Deviation %
Income (families)	
Less than $3,000	0.8
$3,000 to $5,999	8.1
$6,000 to $9,999	11.8
$10,000 or more	– 21.0
Occupation	
professional technical	– 10.3
managers, proprietors	0.6
clerical and sales	– 12.1
craftsmen, foremen	10.0
operatives	14.7
service workers	– 6.5
farm workers	– 3.0
unskilled laborers	6.5
Education (years completed)	
8 or less	4.4
1 to 3 years of high school	– 1.5
4 years of high school	11.0
1 to 3 years of college (post high school)	– 5.1
4 or more years, college (post high school)	– 9.0

ence was often made to a small group of black families who formerly lived outside of town in a "patch" called "Brown Row." Ethnic distinctions were more significant to Coalstream residents. The largest ethnic group was the population of Italian descent, mostly second or third generation. Next to that group in size were residents of English or Scottish backgrounds, called "Johnny Bulls" by some old-timers. The third major ethnic group consisted of families with Polish, Czechoslovakian, and Hungarian backgrounds.

Neighborhoods. Just about everyone in town agreed on the identification and rough boundaries of several residential neighborhoods. The focus was the Downtown, where people believed the bulk of the town's "reliefers" lived, especially on River Road and a few contiguous blocks. The older neighborhoods bordering on the Downtown—Arrow Hill and Pierce Hill—contained long-time Coalstream residents, generally older and smaller families. Mill Hill was similar, but perhaps a little more diverse in the composition of its population. Toward the edge of the town were two more neighborhoods, Summit and Lodi, areas characterized by newer homes, larger lots, and younger families. Parts of these neighborhoods were actually annexed by the town in the 1950s, bringing about a significant increase in the town's population. While Coalstream was a municipality with legal boundaries, Coalstreamers often included other contiguous areas in their definition of their town even though the area was located in the adjoining township. However, this type of inclusion of out-of-town neighborhoods was quite selective. the Brown Row patch, populated by black families on the western outskirts of Coalstream, was not referred to as a part of town. Neither were Iron Hill and Coke Run, in the mountains south of Coalstream, where "hillbillies" lived. But Crestview, a strip of new home development to the northwest, became an adopted neighborhood of Coalstream even though it was really in the next township. The spacious homes with acreage and pleasant vistas made that area a desirable potential addition to the residents' image of Coalstream.

Religion. The ethnic dominance of Italians and Poles made Catholics by far the largest religious group in town. The church, St. Patrick's, previously had an Italian pastor, recently replaced by a priest of Slavic origin. Originally located just out of town on Church Hill, St. Patrick's relocated to the outskirts of Crestview into a modern facility with spacious grounds (part of an old horse ranch). Going to church on Sunday for Coalstream's Catholics necessitated a drive, while Protestants in town could walk to their three churches.

St. Patrick's tended to be quite active in community affairs, particularly under the direction of the previous Italian priest. The Baptists were nonparticipatory. A Catholic priest or a Protestant minister would normally read the benediction or say grace at public meetings instead of a representative of the Baptist church. The levels of regular church service attendance suggested that the majority of Coalstreamers were unaffiliated with any local church.

Lifestyle. Like many small towns, Coalstreamers perceived a division of residents between "respectable families" and other less desirable groups. The criterion used to distinguish the groups was life-style. The respectable class exhibited commitment to the work ethic, maintained a strong family group, and verbalized a strict morality. They frequently contrasted themselves with the reliefers, the low-income families who largely inhabited the downtown area. In the eyes of Coalstream's self-described hard-working respectable families, reliefers behaved differently, making noise and fighting with their neighbors. They appeared disheveled and lived in run-down houses along River Road. Unlike the "respectables," reliefer families were often female-headed households, the fathers having deserted, disappeared, or died. Reliefers were accused of loose morals. They were charged with drunkenness and prostitution.

From the alternative view, reliefers did not have a positive outlook on the so-called respectable families of Coalstream. To them, the working, middle-class families in town showed no sympathy nor concern for the plight of the poor. "Reliefers" often complained that members of more respectable households could be caught drunk in public, engaging in rowdy behavior, and making undue noise, but were not singled out for ridicule and punishment the way reliefers were.

A third behavior or life-style class consisted of retirees, mostly ex-miners and railroad workers. Due to their low pensions and Social Security incomes, they lived the lives of the poor and often appeared little better than reliefers. However, "reliefers were not visible as individuals to the general public, but rather as inhabitants of their dilapidated houses. Retirees and pensioners, on the other hand, were very visible. Lacking anything to do and any place to go, they could be seen sitting on make-shift benches in front of the laundromat or on the steps of the abandoned downtown hotel. Though occasionally for fun or diversion they would call out to passersby, they caused no trouble or damage in town, except for an occasional problem with drunkenness. Even then, drunkenness among Coalstream's retirees brought out remorse, telling stories of their past achievements, bewailing their current circumstances, rather than the more violent behavior of younger drinkers. But their ubiquitous and constant presence Downtown, through which nearly everyone drove, made their's another identifiable life-style in town.

Age. Coalstream was losing its young adults, like many American small towns, due to out-migration for education and jobs. The resulting hourglass population distribution—many at the top (older middle-aged and elderly), many at the bottom (school-aged children), and few in the middle (young adults)—reflects the community's inability to hold or lure back its better educated or more upwardly mobile offspring.

For those who remained in Coalstream, age was a factor of social differentiation. The elderly retirees formed one distinct group, ignored because they were perceived as "dependent." The middle-aged family heads were active in the community's social and civic affairs, perhaps more so than any other age group. Younger family heads focused more on household concerns than on community issues and activities, perhaps because they were less securely established. Coalstream's youth formed a fourth age group, not considered by older residents worth consulting even on issues affecting them directly, such as recreation or community center activities. Teenagers were viewed as troublesome and in need of discipline and control.

Sex. Men and women in Coalstream were relatively equal numerically—but only numerically. In almost every other way, sex was a measure of differentiation, the more important of the distinctions are as follows.

(1) *Age cohort:* Women, 25 years or older, outnumbered the men by 15 percent. Of the population above the retirement age of 65 they comprised 30 percent of the population. This phenomenon may have occurred for two reasons. One is that Coalstream's men tended to be more mobile and more likely to relocate. The other reason is that

women tended to outlive their husbands, leaving many single elderly women in the population.
(2) *Families:* There were seventy families with a single parent. Of these, 80 percent were female-headed households.
(3) *Employment and Income:* Men and women in the labor force earned unequal amounts. The per capita income for men was $5,562 in 1970, compared to $2,465 for women. The reasons for this are revealed by the occupational distribution. Although men and women are equally represented among white-collar workers, most of the men are professionals and managers, while 79 percent of the women work in much lower paid clerical jobs.
(4) *Poverty:* The low per capita income of women contributed to the economic distress for those living in Coalstream. The tenuousness of female employment further complicated their problems, as 21 percent of women were unemployed or laid off for over six months during a year compared to 9 percent for men. While 12 percent of all Coalstream families had incomes below the poverty level, 67 percent of these families were female-headed.

Geography of Social Differentiation

The discussion in Chapter 3 suggests two elements of the geographic or physical expressions of social difference. One is community form, the other is de facto land use. Both indicators relate to the way people actually use the physical features of their community, rather than the ostensible function for which they were intended and built.

Community Form

(1) *Paths.* Paths are avenues of movement for people in a community. Certain ones have particular meaning for residents. In Coalstream, this was particularly true for the county road, which led to the Crestview homes, to the highway, to work and schools (the community college), to the county seat and other cities. The county road was an *exit out and up,* a route to potential betterment. For many communities, Main Street represents the hub of community identity, the center of activities, exchange and commerce. In Coalstream, it leads people past segments of the population many would rather pretend do not exist. Main Street in Coalstream is a definite entry route into the heart of the town's social structure, to be traversed as quickly and as blindly as possible.

(2) *Edges.* Physical phenomena served as boundaries for Coalstream both internally and externally. It was believed by town residents and outsiders that the steep hills around town cut off and isolated Coalstream from other communities in the county. However, social attitudes can overcome hard and fast geography. Although located on a road as steep as that which led to the hillbilly area to the south, Crestview was too desirable to be "edged" out of the residents' functional view of the town. Stronger edges operated internally. The creek running through the center of town effectively cuts off River Road from the town's only park, children's recreation area, and ballfield. Bounded by the creek and, on the rear, by the sharp hills leading to Mill Hill, River Road has only one outlet, at the top of the street, where a reliefer walking to any place in town could be readily spotted and watched.

(3) *Landmarks.* For such a small town, Coalstream had a surprisingly number of meaningful landmarks. Downtown, the old Coalstream Hotel symbolized several things to residents. An imposing structure, it signalled to many Coalstream's prosperous past and current decline. Old timers could remember it bustling with railroaders and businessmen during Coalstream's boom days. For the retirees, it was a place to go with a full day to kill, where they joked and drank with the eighty-year-old owner of the place, Mrs. Ragno. For respectables, it was a physical and social eyesore, a vivid symbol of what Coalstream had sunk to, populated with an occasional part-time prostitute working under the tutelage of the grande dame herself. Abandoned and neglected except for Mrs. Ragno's coterie, the hotel represented to some a site for potential reuse and revitalization, the place where a new Coalstream could start The Coalstream Project set up its offices there, discovering only later that the hotel's reputation as a "house of ill repute" would deter people, especially women, from visiting the staff. Other landmarks included the new church built for St. Patrick's and the old elementary school up main Street in Summit. Both were replete with feelings of Coalstream's steady loss of focus and control of its identity. Centrifugal forces acting upon the town had already caused jobs, populations, and new homes to relocate on the fringe of town. These forces also found concrete expression in the new church—built on the highway and resembling more a shopping center than a place of worship—and in the scheduled

closing of the old school, the last remaining educational function within the town's municipal boundaries.

(4) *Nodes.* The hotel—and Laundromat—meeting places for retirees and reliefers, served as nodes for one part of the population. The Italian Club and the fire hall served a similar function for the opposing social group. Other potential nodes were effectively closed. The community center, rectangular cinderblock structure built by a fraternal club, might have been a node for children, especially adolescents and teenagers, were it not kept chained and locked for fear that its intended users might vandalize the property. The swings at the park next to the creek were also chained, mostly to prevent the nearest children, the reliefers' offspring, from playing—with the same threat of destruction. The ballfield was off limits to everyone but Little Leaguers, due to the town having relinquished its control of the field. Teenagers congregated at the parking lot behind one of the grocery stores. By and large, however, Coalstream offered few nodes for much social activity, the home or places out of town filling the void for most people who chose not to drink at the club or stand around the trucks at the fire hall.

(5) *Districts.* The analysis of neighborhoods as a factor of social differentiation includes identification of areas which are tantamount to districts. Few, however, served as much of a social purpose, positively or negatively, as the Downtown. As the location of the town's lowest rent properties, transients and very low-income reliefers could be effectively channeled toward shelter there and thus checked from spreading into other residential neighborhoods. The two landlords of reliefers, Ragno and Kowalski, were often roundly condemned for their small "slums" on River Road and nearby streets, but never forced to upgrade them or tear them down. Also, the small businesses Downtown, a few groceries, the bars and Kowalski's cafe, would probably find a large proportion of their sales lost if the reliefers were forced to leave. Others in town could drive to the shopping centers and supermarkets along the highway and at the county seat, while reliefers had little option but to shop in Coalstream. A food stamp economy generated by reliefers helped what little commercial enterprises remained in Coalstream survive.

Public and Private Land

As determined by actual use, Coalstream contained very little public land. Most properties, even those legally public by virtue of their municipal ownership, were used by different segments of the population at the discretion of various individuals and groups. The ball field, as previously noted, was Little League domain. Another ballfield located in Lodi was actually owned by a private individual who, it was said, would deny the community use of the land if undesirable people and rowdy behavior appeared. The park Downtown and the community center were locked to prevent use by the teenagers and reliefers. The town's municipal offices were frequented by reliefers only when arrested, knowing that they were quite unwelcome there for any other purposes in the eyes of the community leaders. Mrs. Ragno maintained some influence by owning the large lot Downtown used by almost everyone for parking, a privilege she could quickly revoke if it were to suit her needs. The Italian Club and fire hall were off limits to nonmembers, most of whom were reliefers and other behavioral nonconformists.

Prevailing Modes of Interaction

The operating relationship between the various social groups in town should be described. To do so it is important to identify the major groups and then explore the various interactions between them.

There are five prominent social groupings in Coalstream. One is the blue-collar workers who are employed in nearby manufacturing firms. The second is made up of businessmen and managers. The third is a set of political functionaries operating at the town, county, and state level. The fourth is the "immobiles" made up of pensioners—coal miners, railroad workers, and social security beneficiaries. The fifth is the "outcasts" composed of reliefers and mountain folks.

While most of these groupings seem to *coexist* through a positive but low level of polite exchanges, Coalstreamers use other modes of interaction under certain set of circumstances. There is *competition* from time to time between the political functionaries and the managerial group over who shall govern, and they express their policy differences on Downtown parking, for example.

Benign neglect seems to prevail as the dominant attitude toward the immobiles by the blue collar, business-managerial, and politicians, at least in seeing them as a community problem. In other words, if there were to be any solution it should be on an individual basis with each family taking care of its own.

Conflictive relations were evident between the blue collar, managerial-business, and politicians against the outcasts in a number of different situations. There were occasional fist-fights between the mountain folks and those who frequent the local "beer gardens." Continuing hostility characterized the basic feelings toward the reliefers by many Coalstreamers. The "reliefers" were seen as noisy, unruly, and immoral. The reliefers saw the rest of the community as vindictive, unsharing, and insensitive. Reliefers tended to reject the moral lectures of other Coalstreamers, pointing out how many of the latter drank, used prostitutes, and violated the law. That hypocrisy made the behavioral criticisms difficult to swallow, but just the same, reliefers internalized the respectable norm when it came to their feelings toward each other, condemning others among their own set who might be prostitutes, who failed to take good care of their children, who engaged in publicly obnoxious behavior. It appears that, fearing retaliation, reliefers preferred an interaction mode of *coexistence*; they agreed to be tolerated but generally ignored by the better off in Coalstream. The respectables, on the other hand, expressed a hostility suggesting conflictive interaction, perceiving any benefit to or progress of the reliefers as a distasteful threat to their concept of what Coalstream ought to be.

One event dramatically illustrates the hostile relationship, that is, the response of the Voluntary Fire Company to a fire on River Road where the reliefers lived. One evening the fire whistle blew while some twenty men were drinking at the social club. Running to the front door, they spotted flames shooting into the night sky from one of the clapboard houses on River Road. They walked back to the bar to finish their beers instead of rushing to the fire hall. When asked why, they said it was probably a reliefer's home and they would get there in time to save the chimney.

Coalstream functioned without much of a long-term method of productive interaction among its major social groupings. The pleasant image of a cooperative small town does not seem to hold for Coalstream. Even in times of personal tragedy there was selective response. When reliefers were dispossessed by fire, they were taken in by other reliefers or had to leave town to fend for themselves. Only the respectables attended the funeral of the ex-mayor, and only the county social worker and neighbors would attend the funeral of a reliefer.

Summary

The social structure of Coalstream represents a lower-status community based on the income, occupation, and education of its residents. There are some noticeable inequalities within the community. Occupations are distributed hierarchically (see Figure 6.1). The educational attainment of the adults is bimodal; most of the adults have graduated either from grade or high school. No one group—ethnic or age—is exclusively poor, as there are disadvantaged and more affluent among the various groups. Neighborhoods tend to be identified as having differential status attached to them. The relationship between the groups and individuals with different social status varies considerably.

Influence and Decision-Making in Coalstream

In the world of small towns where tradition has dictated that government plays a minor role, it is surprising to come across repeated examples of active—in fact, turbulent—political systems. Perhaps it is just an American pasttime, that is, politics serves as a diversion. But Coalstream demonstrates that there is a structure of influence, power, and decisions permeating all aspects of community life, from the selection of municipal officers and the setting of governmental budgets to the admission of residents to local organizations and the maintenance of a healthy community life. It may be a serious mistake to think of small towns as decision-making or governmental voids. Minimal formal governance may only hide a conundrum of battles and hard-fought decisions. To counter such erroneous perceptions of small towns, Coalstream may be an excellent antidote.

Rates of Electoral Participation

Chapter 5 proposes three indicators of participation in elections. The first is the percentage of the population of voting age.

Coalstream's proportion of population older than eighteen as of 1970 was 62.3 percent, compared to 67.1 percent for the nation in 1972. The second indicator was percentage of eligible voters actually registered to vote. While 73.9 percent of adults eighteen and older registered to vote nationwide, only 64.6 percent registered in Coalstream. However, the 1972 presidential election found the proportion of registered Coalstreamers turning up at the polls close to that of the nation—74 percent for Coalstream, 76 percent for the nation.

Comparisons for other elections have to be made at levels lower than the whole nation. In the last gubernatorial election, 62 percent of Coalstream's registered voters turned out compared to 64 percent for the county. Municipal elections are difficult to compare. Table 6.7 compares Coalstream to three nearby towns, all located in the same county, with similar forms of government.

Across the board, Coalstream experiences about the same level of voter turnouts as nearby communities. However, the particularly low vote for mayor is puzzling when compared to the mayoral vote in the preceding general election in 1969. At that time, 63 percent of Coalstream's registered electorate voted for that office, in contrast to 68, 67, and 78 percent, respectively, for the three comparison communities.

Attendance at Meetings and Functions

Meetings of the town council were unique events in Coalstream. A hotly debated issue among councilmen would appear in the minutes kept by the municipal secretary-manager as two lines describing the issue and the final vote. Meetings were held in a narrow room, with barely enough seating for members, much less

Table 6.7: Percentage of Those Registered Actually Voting for Important Municipal Offices (1973 General Election)

	Mayor %	Tax Collector %	Constable %
Coalstream	33	42	50
Le Havre	55	36	45
Freeman	60	39	62
Van Buren	67	71	71

for citizens with interest in watching the proceedings. It was occasionally suggested that the meeting room contributed to the habitually low public attendance, but an overflow crowd could easily be accommodated upstairs in the recreation room of the fire hall.

Special hearings prior to the establishment of the Coalstream Project, for example, were practically nil. The public was not consulted on many issues, in part because the government did so little of importance to draw much of a citizen turnout. Municipal budgets were routine, taxes stable and low, and special functions, such as redevelopment, nonexistent. Ceremonies and parades were exceedingly few and largely outside the aegis of the government. The American Legion maintained and placed flowers on the war memorial on Lock-up Hill, primarily to fill the void of the government's inattention to the site. The most conspicuous public ceremony was the annual Memorial Day parade, including designation of Coalstream's Citizen of the Year. The Little League had a big parade on the occasion of the dedication of the ballfield. Regular "battle of the barrel" contests between local volunteer fire companies drew crowds of onlookers. But these latter were recreational functions, not symbolic events of political or governmental significance.

Community leaders took the lack of attendance at governmental functions to mean a lack of concern on the part of residents; that is, the council was doing a good job on behalf of its constituents. To others, holding council meetings in a small municipal office, actually the chief of police's office, was a symbolic gesture commenting that citizen participation was unwelcome. Active efforts to stimulate participation did not exist, leaving the clear impression for voters of a cold reception at meetings. When someone would attend a meeting, the interloper would immediately be asked by the secretary to state his or her business. The effect was to dissuade the person from staying for the remainder of the business, and then after the person left to proceed with the meeting.

Few residents tested the openness of council meetings. In part, the motivation of anyone to attend a meeting would be to criticize or bring a complaint. But that would enable to council majority to identify and single out those in opposition. In Coalstream, people were vulnerable to reprisals and sanctions by the municipal authorities. For example, one former councilman, highly respected in town for his honesty and good intentions, was at one time actively criticizing the performance of the town council. Suddenly, the grass sidewalks of his street were no longer being mowed by town employees. In response to his concern for community issues, he as well as his neighbors were deprived of an important public service. He tried to compensate his neighbors by mowing the grass himself. But sanctions such as public service deprivation and exclusion from certain social circles took their toll and he became completely apolitical.

Another deterrent to public attendance was the perception by most people that municipal decisions were not made in council meetings, but rather in private sessions prior to the formal gathering. Across the street from the municipal offices, key town leaders would sit in the back of Franco's bar determining which council agenda items should receive pro forma ratification later at the official meeting. Afterwards, some would return to Franco's to discuss unfinished business and related items. This is "beer garden politics," public decision-making in private social gatherings. Whereas a citizen had the legal right to attend any official council meeting, only through the acquiescence of the key participants could a citizen gain access to the group sitting around the pitcher of beer where public issues were really decided.

Prevailing Influence Pattern

The text suggests three approaches to finding influentials in a community. The first is to see who holds *positions* of possible influence. Franco and Lordi appear to dominate the most important positions in town. Franco's three roles, president of the council, head of the major political party in town (the same one as that which ran the county), and appointed director of a county public housing project, contributed to an image of great power. Later he added to his influence heading the municipal water authority, although he resigned his presidency of the council in order to do so. Lordi, related to Franco by marriage, was secretary to the council and town manager. This not only provided him with the role of recording council minutes, but the important role of setting council agendas, collecting the town's wage tax, and making most of the day-to-day decisions in the town's government. Other councilmen held positions of some influence. Reddish, the machine shop operator, was next in line to be council president, even though he was allied with the opposing political party. Mayor Serra was the police chief, although day-to-day law enforcement was entrusted to Officer Callan. Callan's sometimes uncertain law enforcement career was always balanced by his continued position as fire chief. Councilman De Pasquale, a truck driver, ran the town's public housing—actually a few converted military barracks. Egan was officially designated Coalstream's representative to the areawide council of governments, although he rarely attended their meetings.

Outside the town council, positions of influence were few. The undertaker, Standish, and the telephone company field director, Williams, were active in the Lions and Odd Fellows, but both groups were rather ineffective. Two other governmental positions—the justice of the peace and the state assemblyman—were held by Coalstreamers, but their involvement in community issues was rarely evident. The same could be said for the head of the consolidated school board, whose activities, now that the elementary school was closing, were relatively minimal and certainly external to Coalstream. On the basis of positions, Franco and Lordi certainly held the most authority.

The next approach is the *reputational.* To learn who were respected as individuals with leadership potential, the staff of the Coalstream Project distributed a survey which asked, "Would you suggest three individuals who have very interesting ideas about how to deal with (Coalstream's) problems?" This question served to uncover people believed to be important, other than those in formal positions of influence. From the responses tabulated, received from approximately one-fourth of the town's households, the town's positional leaders were not always identified as those with the best ideas for Coalstream. The list of those most frequently mentioned revealed the following.

(1) Standish, the undertaker, received by far the most mention. A quiet, retiring man, he was respected for his intelligence, erudition, and proper manners, and perhaps for his ability to advance himself—he made enough money to build a luxurious home near the Crestview area. His popularity was a factor of his adroitness in avoiding controversy. One out of every five respondents mentioned Standish.

(2) The dissident Woods ranked second, even though he had not been active in community affairs for a long time.
(3) Franco, the council president.
(4) Ford, the state assemblyman.
(5) Lordi, the manager-clerk, owner of a grocery store.
(6) Horn, the leading groceryman.
(7) Councilman Egan.
(8) Farrell, Mill Hill leader.

The third technique is *decisional,* identifying the key participants in a number of major community decisions. The names which kept reappearing suggest those who were influential in community affairs. Public decisions are few in Coalstream. Below, each major decision and its main actors are identified.

(1) *Sewerage.* The state's requirement that the municipality install a sewage collection and treatment system in the downtown area was resisted by Coalstream due to the costs it would impose. Lordi served as the town's principal spokesperson, backed by Franco and the council.

(2) *Parking.* Trying to lessen congestion on Main Street, some residents fought and temporarily succeeded in eliminating on-street parking. Franco and Lordi, protecting their downtown property, organized and reestablished on-street parking. The ex-mayor, supported by Farrell, an active member of a St. Patrick's social service group, led the group supporting off-street parking and making Main Street one-way.

(3) *Asphalt Plant.* A proposal to build an asphalt plant on some vacant land near Mill Hill was rejected by vigorous neighborhood opposition, led by Farell, and strong support from council leaders such as Franco and Lordi.

(4) *Flood Wall.* After a major flood, a small wall was built to hold up one bank of the creek, which subsequently collapsed with the next rain. Lordi and Franco, as municipal authorities, played leading roles in negotiating for the wall's construction.

(5) *De Pasquale.* The truck driver councilman who ran the housing project allegedly was caught using rent receipts to pay his own utility bills due to some temporary personal financial problems. Although he fully intended to return the funds at a later date, his ouster was engineered by Franco with the support of the council and Lordi.

The decisions involving the allocation resources and benefits (the asphalt plant, the flood wall, and so on) or the maintenance of the rules of the political game under threat of change (the De Pasquale incident and the like) reveal repeatedly the influence of Franco and Lordi. Although others appear to raise scattered issues, especially Farrell, the Franco-Lordi combination regularly dominated.

Examining the indicators discussed above—voting, attendance, leadership—one can create a composite picture of the patterns of influence in Coalstream. Figure 6.2 depicts Coalstream's political structure. As described in Chapter 5, this is a monolithic community decision-making system. At the top are the handful of leaders who dominate the decision-making power, and their friends and relatives, the "ins." However, note that another group, called "outs," also appears at the top of the pyramid. Outs include people such as Reddish, Farrell, perhaps even Officer Callan, and Williams. Though not currently in the inner circle, they might have been in the past and possibly might be in the future. Although they may criticize the ins such as Franco and Lordi on a personal basis and on political style, or complain about the incompetent management or worse (for example, Reddish called the Franco-dominated council "as crooked as a dog's hind legs"), their own policies when in power did not and probably would not change the outcome of events nor the power relationships in town. All believed in low taxes, a narrow scope of governmental services, and minimal change—values to be explored more fully later in this chapter. They agreed on the issue of the reliefers as well. Without significant alternatives in philosophy or ideas to choose from, voting for ins and outs becomes an exercise in maintaining the town's governmental and decision-making continuity and inflexibility. The vote is therefore not a mechanism of political choice, but a ritualistic or ceremonial legitimation of the rules and output of the established order. That perhaps further explains why so many of Coalstream's eligible electorate choose to abstain from the process.

Beneath the elite is a group of active supporters who can be mobilized not only to vote but to campaign for their candidates and issues. A much smaller group exists that is known as critics—

former councilman Woods, the principal of the high school, and the mailman who combined from time to time to comprise a handful of critics in Coalstream. Their attentiveness to public affairs and their sharp criticism of the council's activities caused most ins to effectively ostracize these critics from many parts of the community network. The sidewalk mowing incident involving Woods is comparable to the black-balling of the mailman from joining the volunteer fire department. At the Italian Club, where the mailman was still allowed admittance as someone more respectable than a reliefer, he would be ignored by almost everyone except some of the pensioners. Though potential opposition to the current leadership, critics were for all practical purposes excluded from the political system. Their latent constituency, the general public, would have been difficult to mobilize in any case.

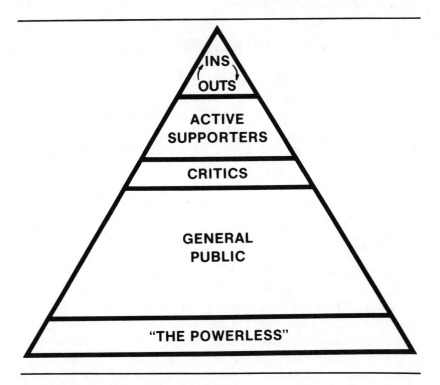

Figure 6.2: Coalstream Political Structure

Despite feelings that all was not quite right with Coalstream, some critics gave up politics, preferring to concentrate on individual and family advancement. Disillusionment was strong among the reliefers, who were vulnerable to reprisals—arson, for example. They squelched their dissent by "carping" among themselves and took no action to change their lot. Youth and very old people also were powerless. Occasionally, there were forays of vandalism by youths who thereby articulated their glee in "getting back at" the power elite.

The general public of Coalstream exhibited a belief, conditioned by many years of supporting evidence and experience, that government was naturally unresponsive, arbitrary, conducted for personal gain, and an "under the table" operation. The phrase, "that's Coalstream," signaled not criticism, but resignation.

Control of Community Resources

There is still a wide range of community resources controlled selectively in Coalstream. We have attempted to identify the major patterns.

(1) *Money.* The undertaker, Standish, seemed to be among the town's wealthiest individuals, although the family owning the firecracker plant and the family owning the gas station were also well-off. Their incomes might not be reflected in the income distribution in the second section of this chapter due to the fact that their official residences were located outside Coalstream's municipal boundaries. Furthermore, there are considerable amounts of unreported income from gambling. Despite Coalstream's official designation as a low- and moderate-income community, the sum of $100,000 was raised within two weeks when Reddish's brothers-in-law needed capital for a gold-dredging operation in Wyoming.

(2) *Credit.* The owners of the two most active grocery stores offered credit to individuals and Franco did the same at his beer garden. Lacking local banks, Coalstream's sources of credit were limited to owners of places which sold needed goods and services.

(3) *Jobs.* The owners of the firecracker company employed the largest number of Coalstreamers locally, followed by Reddish's machine shop. But Franco's control of hiring at the municipal water authority and Lordi's control of municipal employment and

local access to jobs at the nearby glass plant should also be considered.

(4) *Land/Property.* In checking the tax rolls, Reddish, Standish, and a few others were found to be the wealthiest landowners in terms of assessed property values, but Kowalski and Ragno owned the largest amount of property in the Downtown area.

(5) *Information/Mass Media.* Without local newspapers or radio stations, the key possessors of information were several local residents: Lordi, by virtue of his role in collecting the wage tax (thereby knowing the financial situation of everyone in town); (2) Ragno, who due to her vigilant monitoring of comings and goings from her perch at the hotel's front window, knew a lot about the people Downtown; and (3) the bartender at the Italian Club, who also heard a lot of gossip from those who spent their evenings at the Club.

(6) *Social Status.* Acceptance in town was conferred by three key organizations. The Volunteer Fire Department was the most important for male adults. Farrell's committee at St. Patrick's conferred a moral legitimacy to its members. Finally, the Lions, personified by Standish and Williams, reinforced the propriety and sobriety of honest businessmen.

(7) *Access to External Leaders.* Franco's relationships with county political leaders and Lordi's contacts with county administrators constituted the bulk of this limited resource in Coalstream.

(8) *Knowledge and Skills.* Three residents of Coalstream were teachers in the local school system. But there was little evidence that their skills were being used for community affairs except that one served on the town council.

(9) *Personal Qualities.* Several persons commanded popularity and respect from Coalstreamers, but on different bases: Standish, respected for his education and manners; Franco, respected for his power; Woods, respected by some people for his different ideas; Egan (Senior), respected for his willingness to help youth; Farrell, respected as generally a good person; Williams, respected as honest and hard-working.

(10) *Public Office.* Serra, the mayor, did not use the full weight of his office. He deferred instead to Franco, president of the municipal council, and Lordi, the secretary-manager. He even deferred to Callan, the top law enforcement officer. Other public officials were Reddish, second most important council member,

the state assemblyman, the local justice of the peace, and Democratic party committee woman.

(11) *Social Group Support.* Family support was the main social group support in Coalstream.

(12) *Voting and Participation.* Voting was important at election time but of limited influence in community decision-making.

(13) *Leadership Skills.* Franco's control over local decision-making earned him a top leadership role. He had the ability to mobilize people into his camp. Williams, Egan's son, who had been elected to the town council, and the ex-mayor's son were viewed as three persons from the younger generation with leadership potential. Finally, Farrell's ability to motivate people in the asphalt plant and on-street parking controversies suggests a potential for leadership.

(14) *Organizational Resources.* As for organized groups in Coalstream, the volunteer fire department and St. Patrick's church appeared to be the town's most effective institutions. They are able to mobilize significant attendance for dances, bazaars, and especially bingo. Other notable organizations were the Democratic Party, the Italian Club, the Lions, the Little League, and a Parent-Teachers Association (PTA). There were few but occasional linkages to outside regional or countywide organizations, mostly for the purpose of soliciting funds (United Fund) or memberships (community concerts).

(15) *Interpretation of Values.* As will be discussed below, the family and religion are the instruments for translating the respectable life-style of the dominant group into effective community norms. Government, especially outside agencies such as the welfare department with its visiting caseworkers, schools, and highway-situated economic institutions served as important mechanisms also.

The analysis of community resources rests on two questions: which resources are important instruments of influence in the community? how effectively do those who control key resources use them? Not all communities are alike in their valuation of resources. Wealth in one town is the basis of interpersonal respect, whereas in another it may be irrelevant. The student of influence and power in small towns must be keenly aware to detect the nuances of significant resources and the manner in which they are used.

It appeared that control of jobs, social status and personal qualities were important resources in Coalstream. Certain other resources were not very important. The greater wealth of people such as Standish and the firecracker company owners was downplayed, mostly by the owners of wealth themselves, so as not to call attention to their higher status. Access to external leaders was not valued highly, primarily because the regional leaders had so little regard for and involvement in Coalstream. As a low-status community in the county, it could not very well command much attention and compete with more affluent municipalities. the paucity of skills or knowledge was part of its relatively low status. Social group support, voting, and civic participation did not rank very high in Coalstream.

Form of Local Government

Coalstream had established a modified reform local government:

(1) classification of local government: *borough* (much like an incorporated village in other states);
(2) form of government: *council-manager,* with a weak mayor led primarily by the council president and the manager. The council president was elected for a one-year term, in practice reaffirmed every January, while the manager's term was indefinite by default since no one was prepared to replace him on a volunteer basis;
(3) form of electoral ballot: *partisan*—people can vote straight party tickets;
(4) term of the mayor: *four years*;
(5) number of council members: *seven*;
(6) constituency of council: at-large;
(7) term of council members: *four years.*

The council-manager form of government tended to emphasize both the formal and the informal influence of Franco and Lordi. Their respective official positions in the government combined with their effective use of informal resources of persuasion and influence contributed to shaping Coalstream's political system into an elite one. The municipality's major actors, by virtue of who they are, can decide the governance of the town from a backroom table of Franco's beer garden more easily than at the official meetings of the borough council—they are free of the secretary's recordings. In other small towns, informal or extra-governmental power might diverge from formal governmental actors. In Coalstream, formal and informal power appeared highly convergent.

Sources of Municipal Revenue

The revenue data for Coalstream is striking in two ways: its deviation from communities of a similar size and the actual distribution of sources of revenue (see Table 6.8). Compared to other small towns, Coalstream not only received three-quarters as much intergovernmental aid but also only raised one-fourth as much from its own revenue sources. Coalstream's public officials certainly placed a lighter financial burden on its residents than did its counterparts elsewhere in the nation. Half of Coalstream's revenues came from federal and state government sources, while other cities similar in size raised three and a half times more local taxes than they received in outside revenues. The impression of Coalstream's reluctance to expand its scope of government is

Table 6.8: Municipal Revenue in Coalstream

	Coalstream (1973)		Average for U.S. Cities Less Than 2,500 (1972)
	Total	Per Capita	Per Capita
Total Municipal Revenue	$48,048.02	31.26	91.40
General revenue, own source	$24,845.82	$16.17	$70.51
a. real estate tax	$ 7,945.25	5.17	30.72
b. per capita tax	$ 3,832.00	2.49	–
c. wage tax	$11,722.07	7.63	–
d. amusement taxes	$ 345.00	0.22	–
e. charges, fines	$ 1,001.50	0.65	–
Intergovernmental Aid (federal & state)	$23,202.20	15.10	20.89
a. State Aid	$13,147.88	8.55	–
1) cash balance	$ 176.59	0.11	–
2) motor vehicle fuel highway aid fund	$ 9,075.99	5.91	–
3) flood relief	$ 3,895.00	2.53	–
b. Federal Revenue Sharing	$10,054.32	6.54	–
Debt Outstanding	$ 2,500.00	1.63	154.05

highlighted in its virtually nonexistent municipal debt. The town leaders have been unwilling to assume the risk of loans or bonds for municipal improvements and public works. This is all the more important when considering that the community lacks a sewerage treatment system and flood control protection.

It is also important to note how little revenue in Coalstream was generated from property taxes, compared to how much was collected through the instruments of a wage tax and a resident head tax. The property tax burden was only one-sixth that paid by property owners in comparable cities. The wage and the residency head tax raised twice as much as the property tax.

Patterns of Municipal Expenditures

Town government spends virtually all its money on police and highways (see Table 6.9). Revenues exceed expenditures by $14,425. This means that the excess is placed in capital and/or cash reserves to be available later.

Coalstream offers its residents very few services. They receive no public garbage collection, no public sewage treatment, no funds are expended for parks and recreation, not even a contribution is made toward the public library system of the area from which

Table 6.9: Coalstream's Municipal Expenditures

	Coalstream (1973)		Cities Less Than 2,500 Per Capita (1972)
	Total	Per Capita	
General Expenditures (Total)	$33,622.25	$21.87	$91.92*
Highways	$ 7,432.43	4.84	19.67
Hospitals	0	0	.04
Health	0	0	.08
Fire	$ 528.75	0.34	1.28
Police	$10,139.67	6.60	12.20
Sewerage	0	0	2.24
Sanitation	0	0	1.11
Parks & Recreation	0	0	.87
Housing & Urban Renewal	0	0	NA
Libraries	0	0	.25
Interest on Debt	0	0	6.27
Other	$15,521.40	10.09	35.23

*Total exceeds the sum of individual items because capital outlays are included in the total but not in specific items.

Coalstreamers could borrow reading materials. Coalstream has a government that is reluctant to do much other than perform the police and street functions. At best, its residents were offered a limited caretaker form of government.

Summary

Coalstream relies mainly on a traditional form of local government with voluntary community organizations having a low profile. There is some circulation of elites in borough government. The citizens participate mainly during elections and not much in attendance at council meetings to discuss community problems and possible solutions. The leadership patterns represents a consensual elite which holds a minimum scope of government and a low-tax, frugal debt ideology. Local government provides the traditional functions of police and street maintenance.

Cultural Values and Norms

Thus far a number of implicit values and norms in Coalstream should have become apparent to the reader. Disagreements over life-styles and the strongly held values—neatness, hard work, the sanctity of the family—have been discussed. The apathy and lack of volunteerism on the part of the general public reflects defeat and/or the sentiment "don't shake the boat." To improve the lot of reliefers would take an alteration of the majority's intolerance to their life-style. To open up the political system for greater participation and responsiveness would mean confronting privatism and encouraging the noninvolved to risk disrupting their placid environments by expressing political differences or establishing criteria for administrative effectiveness. At the root of each element of the community of Coalstream is a structure of mutually supporting values.

Values in Community Decisions and Issues

Each event described above reveals the implicit values and norms operational in community decisions. The methodology for investigating the roles of values calls for (1) describing the event or decision, (2) identifying the key participants, and (3) searching for the opinions, attitudes, and beliefs underlying the dynamics. Descriptions of five such events follow.

The On-going Sewerage Controversy

Coalstream has been mandated by the state to install an adequate sewage treatment system before it can undertake any major redevelopment projects. This constraint would pertain even to the conversion of abandoned Downtown brick buildings for residential apartments for the elderly. In time, it is likely that the state will compel the municipality to end the flushing of Downtown waste water into the already dirty creek.

The policy positions on this issue divided as follows.

(1) Young homeowners, such as Woods and Williams, most of whose houses used on-lot treatment systems (septic tanks), believed that sewerage was necessary for the overall improvement of the community. Without it, major community developments would be stymied.

(2) The older leaders of Coalstream, both ins and outs, resisted the state's mandate less over the question of need for sewerage treatment, but more over an antagonism toward outside officials dictating to the community.

(3) The town council opposed the mandate in part to establish the posture of resisting outside officials, but more to express its opposition to an expenditure it believed the town could not afford. The council members viewed the state request to connect Coalstream's sewerage system to a larger regional system as costing more than the town was worth. It would require either a tax increase or a bond issue, or both. Lordi broadly hinted that should the state provide the funds to build the system, then the council would accede to the state requirements.

(4) The general public, for what could be discerned about their attitudes, questioned the necessity of sewerage treatment and its economic impact on the community. A new sewer system might spark community development and bring about substantial changes in Coalstream.

The young homeowners were definitely in the minority on this issue. The prevalent values to resist the sewerage treatment came mainly from the older leaders' *parochialism,* the town council's *low tax* position and its desire to keep a *minimum scope of government,* as well as the general public's opposition to anything that would bring about great change.

Traffic and Parking on Main Street

One of the issues sparked by the Mill Hill residents was a concerned for Downtown parking and traffic congestion on Main Street. Led by Farrell, they proposed to eliminate on-street parking and convert Main Street to a one-way street. Opponents who preferred on-street parking suggested instead that the sidewalks be cut back to accommodate traffic better.

(1) The Mill Hill crowd and their allies, again the young homeowners, believed that the crowded and blighted Main Street could be a more pleasant thoroughfare by eliminating on-street parking.

(2) The opposition was led by Franco, whose motivation was economic self-interest; i.e., his trade would suffer if his customers had to park half a block away. He eventually won after an initial setback. The support he garnered was based on a value that stated that nothing should be done to harm small business in any way. In other words, this decision represents the dominance of *small business entrepreneurialism* in public affairs.

The Mill Hill Asphalt Plant

When a vacant lot in the Mill Hill neighborhood was proposed as the site for an asphalt plant, controversy erupted. Although a new factory could increase the municipal tax base and bring new jobs to town, it could also bring the side effects of air and noise pollution as well as truck traffic to a residential neighborhood.

(1) The town council generally supported the proposed development as beneficial to the economy of Coalstream.

(2) Led once again by Farrell, Mill Hill residents and others around the town believed that the traffic and pollution would have a negative impact on a quiet, secluded residential neighborhood.

After calling in consultants, such as a biology teacher from the local high school, and taking the firm to court, the development was defeated. The successful resistance to the proposed industrial intrusion represents the values of *neighborhood preservation.*

The De Pasquale Ouster

When Councilman De Pasquale was charged with the misuse of public housing rent receipts to pay his personal utility bills, the issue was hushed up and quickly resolved. Kept within the town council, it did not become a controversy involving other Coalstream residents. Allegedly there were numerous other sorts of public corruption in the form of kickbacks on public works. The De Pasquale affair involved the use of public dollars for private interests. De Pasquale initiated the issue, immediately followed by his offer to repay the $40 or $50 involved. Franco and Lordi expressed their righteous indignation. De Pasquale's removal was sought from the time of his election. In this case a clique member was penalized for lack of *conformity*.

The Closing of the Elementary School

The decision to close Coalstream's last vestige of the once locally operated school system was made outside the community by the regional school board with the willing acquiescence of the town's one school board member. Thus, the closing was presented as a fait accompli, with virtually no one from Coalstream appearing at the public hearing. The issue, therefore, was not whether the school should be closed, but rather how to adjust to the closing. The ramifications of the closing were significant. As some individuals remarked, the attrition of local educational institutions vividly represented the waning of Coalstream's ability to provide for its citizens. Educational norms would no longer be those held in Coalstream, but those of the outside world.

The issue then became an adjustment, how to reuse the abandoned building in face of the likely negative impact of an empty school in a residential neighborhood (Summit). The interests divided up as follows.

(1) The Volunteer Fire Department expressed interest in converting the school into a new fire hall. It would move from the present Downtown building with its tortuously narrow exit.

(2) The town council suggested that it be torn down to make room for a park.

(3) The consolidated school district hoped to recapture its investment by selling it commercially or reusing it for other educational purposes.

(4) A few in town, led by the senior Egan, thought that the solid structure might be converted into housing for senior citizens, although the distance elderly residents would have to walk to get to downtown grocery stores made that idea a bit impractical.

(5) The general public by and large complained about the decision, but absolutely no one came forth to mobilize the opposition beyond citizen protest at a few school board meetings.

At the date of this writing, no decision has been reached about the property. The general public's passivity seemed to dominate. As a value, it might be considered part of the community nondecision approach and a part of the feeling described as "let nature take its course." This inaction leaves another abandoned building in town.

Each of the five community issues discussed above highlights central elements of the value structure of Coalstream. Figure 6.3 depicts both their community values and norms. The general impression is of reluctance to engage in anything other than minimal change in the face of new proposals and opportunities. Arguments for housing, economic development, improved public decision-making gain acceptance if they fit in with these values and norms.

Institutional Roles in Values

The analysis of the five community issues above shows certain participants were quite active while others passively watched from the sidelines. Certain institutions can be identified easily as "outside the fray."

Figure 6.3: Dominant Coalstream Values and Norms

Issue	Community Values	Community Norms
Anti-sewerage	Parochialism, low-tax minimum government	Non-compliance
On-street parking	Small business profits	Business dominance
Asphalt plant	Neighborhood preservation	Protect residential neighborhoods
DePasquale ouster	Conformity	Acquiesce in the face of political power
Reuse of abandoned school	Let nature take its course	Non-decision

(1) *Economic Firms.* Other than Franco's own economic self-interest in protecting his tavern, there is little evidence of economic firms playing much of a role in shaping community values. The asphalt plant failed in its attempt to locate in Mill Hill, and likewise failed to energize other businessmen to rally to its cause. The Downtown grocerymen seemed determined to watch community affairs from the sidelines. External economic firms appeared to acquiesce to the norms of Coalstream by minimizing direct involvement in community affairs and cooperating with local elites.

(2) *Schools.* Now very much an outside agent, the school system maintained a "hands-off" attitude toward Coalstream. Even local representatives—a school board member, a school principal, and two or three teachers who resided in Coalstream—minimized their involvement in community issues. If the school system offered opportunities for shaping new values, the effects were exhibited by the children who generally found it necessary or advantageous to leave town upon graduation. The relatively low status of Coalstream compared to other communities in the county might have contributed toward the prevailing attitude in the schools that Coalstreamers were somehow not as capable as other students.

(3) *Religion.* Despite the great drawing power of church activities, such as St. Pattrick's bingo, and the frequency of people citing the importance of religious belief, there was little evidence of direct church involvement in community affairs. The new priest seemed to have no position on any of the issues discussed above, nor did Farrell's layman service group at St. Patrick's show interest. The other churches similarly were not involved. The previous priest at St. Patrick's seemed to have been more activist-minded, but he was eventually replaced. Churches taught the values of morals and religious belief, which were expressed as cleanliness and hard work, and adopted by most families, but without corresponding church values of community responsibility or philanthropy and compassion for the downtrodden and dispossessed. Coalstream showed little evidence of being a compassionate place.

The stronger institutional generators and protectors of values seemed to be the government and the family.

(1) *Government.* The dominant community values were mostly vociferously expressed and defended by the town council, the

only significant government actor in Coalstream. Decisions engineered to conform to dominant norms originated in the council chambers, or perhaps more accurately, in the back room of Franco's bar. Reprisals or sanctions against those deviating from community norms were carried out by governmental actors. The grass along the sidewalk would go unmowed, or individuals would be arrested; the various sanctions could be quite effective and devastating. The rate of change and the types of change allowed to surface in Coalstream were carefully regulated by the council. Overall, in this small town government served as the protector of community norms and values.

(2) *Family*. In Coalstream, the family appeared to operationalize values into norms. The community norm of slow, traditional change was an outgrowth of and bolstered by the family's inward-looking privatism, its concern for its own problems, and its willingness to let community problems take care of themselves. The community norm of ignoring or even punishing those who deviate was enforced on a daily basis by the family norm of conformity to a respectable life-style and intolerance toward behavioral deviants such as reliefers. Government's shaping of community norms found the family's guardianship of individual and interpersonal values a necessary and supportive match.

Community Priorities

The final consideration is the difficult task of determining community priorities. Extensive efforts were made with door-to-door surveys and neighborhood meetings to ascertain Coalstream residents' hopes, desires, and expectations for the future. It was apparent that the community's stagnation was not really the result of residents' complacency. Three-quarters of the respondents to the community survey conducted by the Coalstream Project complained about conditions Downtown (see Figure 6.4). Nearly half bemoaned the lack of recreational facilities and activities, especially for teenagers. Similar proportions of respondents mentioned problems of law enforcement, police protection, poor infrastructure (sewerage), and the lack of jobs and commerce. Neighborhood meetings brought forth proposals for senior citizen housing, a roller-skating rink, a skillsbank, movies, and dances.

At about the same time the "politically relevant"—key participants in community affairs—were asked how important specific

Figure 6.4: Concerns Raised by Residents of Coalstream

Law Enforcement
1. inadequate police protection (22)[a]
2. troublemakers and loafers on street corners (13)
3. underage motorcyclists speeding through downtown (5)

Housing
1. dilapidated downtown buildings, "eyesores" (37)
2. inadequate housing quality (21)
3. lack of good housing to rent or purchase (6)
4. insufficient housing quantity (6)
5. unkept buildings (5)
6. vacant buildings (5)

Recreation
1. recreation (46)

Governance
1. poor council leadership (16)
2. apathy, lack of civil pride (21)
3. clean up depressed areas (10)
4. what to do about reliefers (9)

Commerce
1. not enough stores, shops (9)
2. no job opportunities, especially for young people (34)
3. no industry (15)

Infrastructure
1. inadequate flood control threatening homes, development (12)
2. lack of sewage treatment (12)
3. inadequate downtown parking (9)
4. road upkeep (5)

a. The number in brackets represents choice by households (those receiving 5 or more nominations in a survey of 138 households).

solutions were for improving Coalstream. Figure 6.5 illustrates the answers of the three most influential groups in town (see Figure 6.2 for their position in the power structure). There was unanimous support to install a sewer treatment plant, provide flood control, and tear down old buildings. There was little support for changing the railroad that bisected the town, and there was considerable disagreement on the need to build a new town hall and fire hall, with the ins as the major supporters. The outs did not go along with the other key participants in "ridding the town of undersirables." Both the residents and the politically relevant agreed that something should be done to alter the condition of the dilapidated Downtown buildings.

Figure 6.5: Support for Proposed Solutions of Coalstream's Problems

	Ins	Outs	Critics
Unanimous support for:	More retail stores Sewage Flood control New fire hall Tear down old buildings	Sewage Jobs for youth Tear down old buildings Flood control More residents	Sewage Tear down old buildings Fix up buildings Flood control
Majority support for:	New industry Jobs for youth Fix up old buildings Rid town of "undesirables" New borough hall New park Parking Roads More visitors	New industry More retail stores More parking Extend Main Street Change railroad New park More visitors Fix up old buildings	New industry Rid town of "undesirables" More visitors Jobs for youth More parking More residents New park & recreation Change railroad
Even split			
Minority support for:	Extend Main Street Change railroad	Fix up roads New fire hall Rid town of "undesirables"	New borough hall New fire hall Fix up roads More retail stores Extend Main Street
Unanimous lack of support for:		New Borough Hall	

Preferable Future

Ultimately, much more investigation would have to be done to uncover residents' deeper aspirations for the future of Coalstream. But certain themes and images of desirable future states of affairs can be discerned. Two models were mentioned repeatedly in a positive way. One was Harmonyville, a town in the northern part of the county, a place of considerable social activity. Parades and fairs drew countywide attention and newspaper coverage. This community projected a sense of resident pride in the quality of community life. Hunters Crossing might serve as the second model. Situated in the southwest part of the county, this community enjoyed a thriving economy despite its small size. A bustling downtown providing comparison shopping and restaurants for residents and visitors plus nearby, clean factories combined to create a pleasant picture of the community.

Coalstream might have generated an acceptable image of the future for residents to plan and act upon if they would explore these communities in more detail. Harmonyville had a reputation as a clean and orderly residential community with good housing, abundant recreational opportunities, and strong community feeling and pride. If Coalstream followed this model it would develop a suburban residential existence with shops and stores, and employment would be left to regional centers such as the county seat and other communities capable of generating economies of scale. The alternative of Hunters Crossing is more one of a complete small city, providing a full range of services and functions for its residents. For Coalstream, this would involve redeveloping the commercial and employment activities it had before the collapse of the coal industry. Business expansion, perhaps industrial development, and greater governmental provision of infrastructure and amenities would be called for. By choosing one or the other of these "generative themes" or images, Coalstreamers might then be able to sort out desirable from unwanted community changes. In the absence of thinking about decisional preferences for the future, community change in Coalstream is likely to be hamstrung by the dominance of the norm against change.

Summary

The prevailing community values and norms are itemized in Figure 6.6. Local government and the family are the major stimulators and reinforcers of these values and norms. There is unanimous support among the community leaders for sewage treatment, flood control, and tearing down old buildings. The community as a whole is searching for an answer to its long-range future.

Figure 6.6: Coalstream's Values and Operational Norms

Community Values	Community Norms
(1) sanctity of family life	— intolerance of deviating life sytles — male dominance
(2) low-risk, nonchange	— reluctance to make decisions — "let nature take its course"
(3) individualism and privatism	— minimum scope of government — low taxes
(4) local-autonomy and parochialism	— wary of outsiders — non-compliance with mandates from outside
(5) entrepreneuralism (what is good for business is good for the community)	— business interests dominate
(6) loyalty to group	— ostracize citizens

NOTE

1. The project was to assess the feasibility of strategies for small town revitalization. The design was to go through four phases: (1) develop an inventory of the town and its assets; (2) define goals and programs for renewing the community; (3) understand how feasible each possible program might be; and (4) come up with an overall plan for renewal. A number of detailed reports on the project have been produced by the Institute on Man and Science, Rensselaerville, New York.

Chapter 7

COMMUNITY ANALYSIS AND CONSIDERATIONS FOR DEVELOPMENT

We have proposed community analysis as a way of better understanding small towns. The key elements are values and norms, social structure, the economy, and decision-making patterns. We have not only developed the major concepts of each by reviewing and summarizing the relevant literature on small communities, but we have developed specific factors and operational indicators that can be measured and collected from readily available data sources. Furthermore, we have developed a discussion guide that can serve as the basis for exploring more deeply what people think and feel about their town.

All of the above should help students, concerned citizens, and professionals better appreciate the places they study or their living and working environments. It should also provide a better understanding of a specific community for those about to engage in solving local problems, planning a purposive change, and bringing about some sort of community action. We believe the participants need to know something about social and economic conditions, as well as how the town makes decisions and what the prevailing values and operational norms are as they contemplate community development. This is true whether the action be initiated by a resident or by professionals hired to resolve particular community problems.

Before proceeding it is important to restate and refine our original notions about community analysis as a way of thinking about small towns. Figure 7.1 is a schematic representation of the three basic elements of the community system—social, economic, and political as immersed in the fourth element of cultural values and norms.

Each component has been treated separately in Chapters 2 through 5 and applied in Chapter 6. We have also identified some

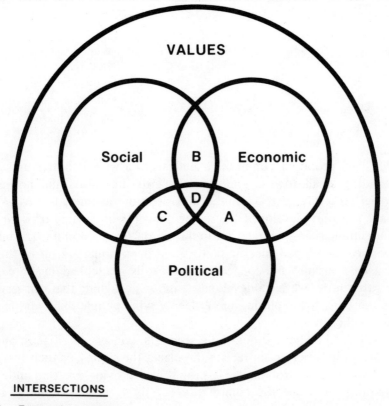

INTERSECTIONS

A - Economic - political
B - Social - economic
C - Political - social
D - Inner - core

Figure 7.1: The Four Elements of Community Analysis and Their Intersections

of the connecting values; that is, the concern that the local economy is sufficiently profitable to attract economic firms to provide jobs and resources to improve the quality of life. There is also the concern that the political decision-making process operate in such a way as to provide an orderly if not consensual way to resolve the community's character and direction. Similarly, the social values are expected to balance the paradoxical relationship between achievement and equality as part of the American experience.

In some communities these four components may be highly integrated while in others they are not. That is, in a highly integrated town those who own or operate the economic firms are most likely to dominate the local government, live in the most prestigious neighborhood, and shape and maintain the prevailing values and norms. In this regard, Presthus states that small towns have greater overlap than larger cities. He draws this conclusion in comparing his study to the study of New Haven, Connecticut, a city of 150,000 people, where Dahl found only a 6 percent overlap of leaders engaged in more than one of three issues pertaining to schools, urban renewal, and political nominations.[1] In his own study of two small towns under 10,000 he found 32 and 39 percent overlap.[2] Presthus measured the overlap of elites who participated in community decisions on a school bond issue, hospitals, new industry, flood control, and municipal building. Dahl also measured the overlap of memberships of persons attending the social cotillion, being on the board of directors of sizable economic firms, and holding a position in political organizations or city government, and found a small degree of overlapping membership.[3]

Given the greater probability that small towns are highly integrated, although not all are, an effort should be made to determine the degree of integration. Here we believe it important to indicate some of the relationships between and among our four elements that are identified in the intersections of Figure 7.1.

In A, the *economic-political* intersect, the people holding considerable economic resources—money, jobs, credit, and the like—are more able to use them to determine who holds public office and what local government should or should not do. They are most likely to see that government protects their interests or at least does not jeopardize them, whether it be through tax policy, land-use regulation, or public services. Some economic dominants do not enter the political realm directly because their interests are well served by the town council or because local government is not likely to be doing anything that negatively affects their interests.

Local public officials, on the other hand, relying on their popularity or ability to win elections, may intentionally or unintentionally affect the local economy through town hall decisions. Of course, they may be members of the economic elite as main street merchants, or be well socialized to do what is best for business. Others may be blue-collar workers who seek an opportunity to get back at management and correct the "wrongs" that occur each day at the workbench. While most public officials favor proposals to boost the local economy, few know what really can or should be done to make their town more competitive in attracting economic firms, and therefore most local governments leave economic development to others. The prevailing ideology on the politics of the local economy in small towns is generally supportive of the marketplace, its preferences, and operations. In Coalstream, for example, the low-tax, individualistic parochialism or commitment to local autonomy has produced not only a minimal scope of government with limited public services, but also a place unattractive to outside economic firms.

In B, the *social-economic* intersect, the people with the highest social prestige in the community may have attained that distinction as a result of their economic resources, and therefore prestige and economic resources are closely linked in the socioeconomic status positions held in town. In fact, there is a very high correlation between income, education, and occupation consistently found throughout America. Form and Miller, in reviewing many community studies, assert: "Since industry contributes a large part of the occupational composition and differentiation to a community, its influence in shaping social classes is very great."[4]

The connection may be more likely and visible in small towns than in big cities as the use of social groups as a way to rank individuals and families is more generally known and intimate. Not all persons fit well into their status position, however, as newcomers or those with new wealth may find it difficult to gain social acceptance. In Coalstream, for example, those who owned land or businesses were perceived as of a higher status as long as their life-styles were considered respectable.

In Coalstream, acceptance by the male-dominated society was a prerequisite to attaining public office. Holding a public office, in turn, enhanced a person's prestige in the community. Group loyalty was also tied to remaining in local office, and breaking the norms of sharing was cause enough for eviction.

In C, the *political-social* intersect, people holding public office and/or authority also have high social status by virtue of their positions in prominent community organizations. Such people generally have gained a reputation and experience in demonstrating their concern for the community. They have also developed leadership skills and are able to garner social group support and to mobilize organizational and other community resources. Therefore, to assume public office is a "natural" progression for such people—to extend their scope of influence from the confines of social clubs and associations to the broader aspects of community affairs. For that matter, the means and ends of community organizations are generally perceived and treated as though there is little or no difference from those tools and skills used in local government. In Coalstream there was an obvious distinction between the relatively powerful and the powerless. The former was embodied by the "political boss" who combined local authority, economic resources, and social status. He also held and reinforced the prevailing community values, norms, and ideologies. The latter, reliefers and pensioners, had little political influence, economic resources, or social status, and generally lived a life-style considered unacceptable to the majority of the community.

D, the *inner core,* is the center where all four elements interact. Here may be people with considerable economic resources, high social prestige, great influence, and embodying the prevailing community values and ideologies. This is likely to occur in the one-company town, especially if the owner-operator is a local resident. There may be other people at the other end of the scale with few resources, low prestige, and little influence.

Considerations for Community Development

Our framework of community analysis has alerted us to a number of factors that should be considered by those who wish to engage in community development and change. They are discussed here as cautions that require some examination as action programs or projects are contemplated. We have identified five areas that should be considered:

(1) the degree of penetration of a small town by external forces;
(2) the priorities of community needs and problems;
(3) the profiling of community conditions;
(4) the purposes (ends) of community development;
(5) the approach and/or strategy and tactics (means) of community action.

The Penetration of the Small Town

It is an important consideration for community development to know to what degree a small town has already experienced various kinds of external influences upon its way of life. It is also important to know the degree to which the community has or has not insisted on retaining its own identity and how it has attempted to do so.

Predepression studies viewed the small town as an autonomous and permanent fixture of the American landscape.[5] The studies of the 1940s noted how the small town was drifting away from the swiftly changing mainstream of American life.[6] In fact, two main concerns guide most small town research and development since World War II. One is that the small community has been totally penetrated by national forces and all but absorbed into the larger mass society destroying its identity, tradition, and the spawning ground of most Americans of the past. The second is that the isolated and insulated ruralite—farm and nonfarm—and small community are being effectively linked to the mainstream of American life.

Some analysts view with alarm the destruction of the independent small town as they become subordinated to the forces of urbanization, industrialization, and bureaucratization.[7] Others applaud the demise of small towns because they are believed to be essentially antidemocratic.[8] Still others are unaware of or ignore the penetration thesis and continue to study and act as if the community were autonomous or to propose regional plans to coordinate these localities into the larger society.

Other analysts support bringing greater opportunities, conveniences, and amenities to those left out of the consumer society.[9] While some are sanguine, others are busily constructing public and private networks which link small towners both vertically to the national scene and horizontally with regional coordinating mechanisms. In any event, the point is that whether one does or does not approve of the penetration of the small town, it is happening. Martindale and Hanson have noted that small towns have begun to lose ground with many vanishing altogether.

> In the twentiety-century world the Jeffersonian ideal of autonomous small towns has become an anachronism. Power is shifting from locality to the great centers of government, industry, and finance. If the small town survives at all it is not as an autonomous center of local life but as a semidependent agency of distant power centers.[10]

The above authors believe the drama of local life is the "clash of the old ideals of the autonomous community with the realities of survival in a mass world."[11] Vidich and Bensman reinterpreted their earlier observation on the loss of autonomy to note a historical confrontation in process where they anticipate an "enduring battle between the proponents of populist culture and the bearers of the new urban life styles."[12]

Warren has identified four ways American communities may differ.[13] One is in their *local autonomy,* the extent to which a community is dependent or independent of extra-community units in exercising control over its own decisions. That is, what control does a community have if an absentee-owned company discontinues its branch plant, or the state highway department decides to build a new highway through town. The second is the *coincidence of service areas,* the extent to which stores, churches, schools, and the like provide services based on a common boundary easily perceived as being in one community. That is, do services coincide so that everyone within the community is served by institutions within the same community, or does a family live in one school district, attend church in another locality, and trade in still another without any common center of community activities for that family. A third is the *psychological identification with a common locality.* That is, do the local inhabitants consider the community as an important reference point, or is there a weak sense of identity with little feeling of "belonging" to a particular community. Finally, the *horizontal pattern* indicates the degree to which the various local units—individual and social—are interconnected by common sentiments and behavior.

Warren believes that the loss of community autonomy and identification with the community—apathy, alienation, and anomie—constitutes barriers "to the efficient mustering of forces to confront community problems."[14] He proposed that community development is one way to overcome these barriers. It provides

> a process of helping community people to analyze their problems, to exercise as large a measure of autonomy as is possible and feasible, and to promote a greater identification of the individual citizen and the individual organization with the community as a whole. Through such a process, communities may be helped to confront their problems as effectively as possible.[15]

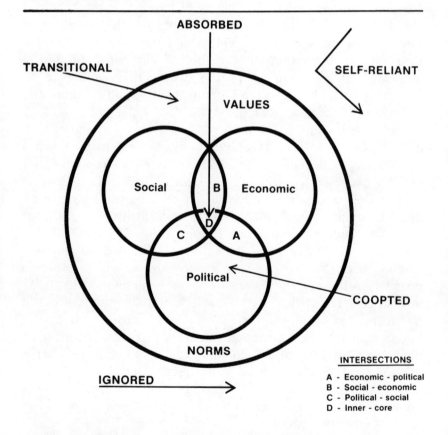

Figure 7.2: Degree of External Penetration of Small Communities

Using the community analysis framework, we discern three types of more or less penetrated small towns and two types of relatively autonomous communities. Figure 7.2 illustrates the five kinds of places designated by the direction of the arrow of external forces that potentially bear on small towns.

The first is the *ignored* community. The residents of these towns have been left out of the benefits of the national prosperity. They exist day to day without much support or involvement with the mainstream of America. What support they do receive tends to be token gestures of help from designated external agencies mandated to provide services. In a Senate hearing on rural development, an Iowa senator complained about the failure of the Farmers Home Administration to help the neediest small towns in the southeastern corner of his state. "It seems to me the greatest problem is that there is no outreach at all. People in those areas are not being served."[16] Residents of these towns are at a disadvantage by an apparent lack of options for improving their conditions. They are offered no assistance by neighboring towns or other governmental units. To the extent that they have been ignored, they have been "written off" as beyond hope. Frequently, residents of these places see themselves as stuck in a rut, with no place to go. Towns that were once mining or one-company towns abandoned with the closing of the mine and the desertion of its more energetic residents tend to settle for a pattern of marginal existence. The main changes occur as a result of the continued deterioration of physical and social conditions and the slow attrition of population succumbing to old age.

No one seems to come to the front and take the initiative to improve these towns. However, this does not necessarily make them wide open for external forces to introduce an improvement program or fix-up project. The residents are likely to be suspicious of the motives of those from outside who suddenly have decided to help the community. Trust must be established and based on specific project activities that are meaningful to the local residents. It may even take some time to determine who speaks for the residents of a town of this type as there is probably little consensus on what are the community's problems, let alone what should be done about them. The residents of these towns have little experience in working collectively on community problems and maintain a relatively weak identification with the locality as they watch the rest of the world go by.

The second is the *self-reliant* community which prides itself on having successfully resisted the ways of the larger society. Residents in these towns are quite independent. They feel a strong sense of identification with *their* town and have managed to build a way of life that fits their preferences, develops jobs and services, and provides amenities fostering a higher quality of life for themselves. Their successful resistance to the subtle and pervasive encroachment of the larger society implies a strong sense of community and considerable local control. This is difficult to achieve in an interdependent world with externally supported education and mass media systems. It takes a determined leadership with deeply held ideologies to face the challenges to small town self-reliance. Successful resistance to and accommodation with the many forces that attempt to penetrate also imply particular kinds of local social and governmental structures. In the social realm, self-reliance implies a cooperative mode of interaction among the residents to minimize internal dissention which might warrant the intervention of outside forces. For cooperation to succeed, however, the decision-making process is most likely consensual based on shared community values and norms.

The self-reliant community tends to be fairly satisfied with its present position, making necessary adjustments to changing conditions. Any effort to introduce additional changes generally are resisted. Thus if some splinter group preferred alternatives to the prevailing situation, it should expect considerable opposition. Such a group may have to concentrate on modifying the agenda of local decision-making and/or be prepared to go so far as to replace the existing leadership. To introduce change from outside the community probably will be even more difficult as the town has successfully resisted these forces before and knows how to protect itself. Government in these towns will try to "go it alone," foregoing efforts by other governmental units to participate. It generally perceives incursions by state and federal units as undue interference in local affairs, and will likely resist policies and programs which emanate from outside. If externally induced community change is to be pursued, then the outsiders must gain local sponsorship from the local leaders and carefully demonstrate the costs and benefits of any proposed change.

Third is the *transitional* town which has experienced the generalizing effects of being part of the larger society. The residents have been socialized to the values of the greater American society and its main symbols. The mass media enters the homes of each family projecting on the screen what is happening in the country. The messages might sensitize youngsters to challenging opportunities available elsewhere, or it may project risks and fears which reinforce the belief of some that conditions at home are better than those in the big city where crime, congestion, and conflict appear to be a daily experience. The result of the moderate amount of penetration is that the sense of community values are now more subject to change, local autonomy is only moderate, as is the sense of attachment to the town.

It is difficult to really tell which comes first, whether the effects of the external world have reached individuals or local institutions first in small towns. We suspect it is the individuals, especially opinion setters who, once learning about alternatives to local conditions, select out what they prefer and seek to adapt it to the small town setting. They tend to invite change without full consideration being given to impending consequences. They perceive the short-range implications rather than exploring the ramifications; i.e., that a new economic plant may not only bring to town more people but different kinds of newcomers with differing preferences and expectations.

The opinion setters are the main initiators of change. Their prestige in community affairs is generally sufficient to gain acceptance for their proposals, but they may not be in charge of implementation, and therefore there may be a substantial gap between what was intended and what was achieved. Opinion setters are fairly well disposed to work with outside professionals and experts in community development. They respect the experts' status and believe they can and will resolve the community's problems. One word of caution, however; since the community has little experience with outsiders, not all key influentials may willingly accept the outsider. Nor may the outsider meet the community's needs. The solution to one problem in a transitional community may lead to additional as well as long-range problems.

The fourth type is the *coopted community* where penetration of some of the local institutions has taken place. The leadership in coopted communities have perceived that the community will benefit most by connecting with the mainstream of American life. They have lured the external investment of absentee owned businesses and factories and tapped into the abilities of federal and state agencies to assume many municipal and community functions. These communities, of course, benefit from using the nation's prosperity, but do so at a great cost. Decisions made outside the community and fluctuations in national or regional conditions reverberate through the town. The challenge to this sort of small town is to get what it needs from external sources without sacrificing too much local control or incurring too many burdens. The process of maintaining a dynamic exchange relationship with the outside world requires certain competencies in these towns. Concerned about receiving financial aid from outside governmental units and luring business investment from other places, coopted communities may develop increased professionalism in local government in order to better deal with professionals at federal, state, and regional levels, frequently through the council-manager form of government.

The new managers, or trans-locals as Martindale and Hanson refer to them, develop distinct life-styles and pursue strategies different from the locals. The latter "tend to press the community in the direction of self-sufficiency."[17] The former "tend to orient life to the centers of outside power."[18] Therefore, one might expect considerable polarization of opinion between the two types with the balance of force eventually favoring the new managers who can mobilize and draw from a greater pool of community resources. The newcomers are also more likely to recruit expert community developers to solve problems. These professionals are carefully selected to serve the interests of the newcomers.

The final type of community is the *absorbed* one which has been totally penetrated by external factors and the town has become indistinguishable from population aggregates around it. It may be simply a "spot" in the road servicing those who pass through, a suburb with little sense of community as the commuter uses the place merely to sleep, or a mill town whose sole purpose is to serve the interests of the mill owners. Therefore, the community is quite dependent with very weak attachments by the residents whose interactive pattern is less directed within the locale and more extended beyond the immediate area. There is a real loss of community, as Swanson suggests, where there "is an inarticulated ideology that allows for the atomization of the people in a homogeneous setting."[19]

If there is to be any improvement in the quality of life in the absorbed community, it generally must await the decisions of external institutions and their willingness to deliver services to the area. Locally initiated community action usually starts from scratch. Potentially any and every resident is a participant. If there is local initiative to improve the community, there will be a tendency for some group to assert itself as the spokesmen of the town. As they mobilize for community action, they will tend to take on the mantle of local leadership and determine the agenda of action. If external community developers are called in early enough in the process of revitalization, they may be able to develop a representative basis of community support for action.

Coalsteam, in the 1970s, was a transitional town. It had become desperate in searching for some replacement for its lost economic base. The residents' children were forced to leave town in order to take advantage of economic opportunities elsewhere. The Borough council turned to outsiders for help with ideas, consultants, and money. However, the leadership cautiously protected their own autonomy and were in noncompliance with the state's mandate for sewage treatment. The strong desire on the part of the citizens to tear down the old buildings downtown because they were an eyesore and embarrassment was a familiar American sentiment.

Community Needs and Problems

When asked, the citizens and leaders of Coalstream brought forth a long, wide-ranging list of community problems. (See Figures 6.3 and 6.4 for a list of problems and possible improvements.) The list of improvements, despite the community's cautious and tentative approach toward change, contained ideas for almost every element of community life. It is a rare small town that is totally complacent and satisfied with the way things are. Certainly many, perhaps most, residents may see no reason for a total community overhaul, but there are always some suggestions about what ought to be done to improve the quality of community life.

For that matter, the concerns of small towners are relatively similar to those of many other Americans. Watts and Free found in a 1972 attitudinal survey that in general small towners like the rest of the country were more concerned about domestic issues here at home than international affairs. People throughout the country were most concerned with violence, inflation, crime, and drugs, yet small towners were somewhat less concerned about poverty, education, unemployment, urban renewal, housing, and urban problems than were other Americans.

Local public officials in American small cities (those under 10,000 in population) were very specific in identifying the critical problems in their community (see Table 7.1 study concerns some thirteen hundred cities), two-fifths ranked insufficient municipal revenue most critical, followed closely by the fact that their community does not have enough industry.

Leaders in four small and four large Georgia communities differed, however, on community needs (see Figure 7.3).[20] Small town needs pertained to streets and highways, more industry, zoning, medical facilities and doctors, and water systems. Large cities concentrated instead on governmental reorganization and services, environmental controls and sewer systems, and public transportation. In Coalstream the citizens identified dilapidated buildings, need for jobs, and police protection (see Figure 6.4). Leaders agreed upon the need for sewers, flood control, and

Table 7.1: Rank Order of Critical Problems of Small Cities

	Problem	Cities Number	Percent
(1)	Insufficient municipal revenues	581	43.3
(2)	Not enough industry	475	35.4
(3)	Inadequate community facilities	368	27.4
(4)	Lack of housing	293	21.8
(5)	Insufficient growth in central business district	258	19.2
(6)	Lack of jobs	179	13.3
(7)	Inadequate public transportation	104	7.8
(8)	Too much growth	91	6.8
(9)	Lack of education and cultural opportunities	46	3.4
(10)	Other	141	10.5

SOURCE: Adapted from Michael J. Murphy, *Governmental Data in Municipalities 25,000 and Under* (Urban Data Service Reports, Vol. 7, No. 1, International City Management Association, January 1975), p. 15.

Community Analysis and Considerations for Development

Figure 7.3: Mean Rank Order of Community Needs — 1968–1975

Mean Rank Order		
4 Large Communities	4 Small Communities	Needs
1	20	Consolidate city and county governments/services
2	1	Improve/expand recreational programs/facilities
3	8	Improve/maintain public education
4	12	More/better housing
5	19	Cooperation/unity/communication
6	30	Environmental control
7	15	Improve/expand sewer
8	29	More low-income/rent housing
9		Public transportation
10	11	More jobs
11	9	Better law enforcement
12	13	Improve welfare program
13	23	Reform tax structure
14		Improve government services
15		Improve garbage collection
16	2	Improve/maintain streets and highways
17		Improve long-range planning
18	26	More tourism/tourist attractions
19		Joint city and county planning/development
20	16	Improve traffic flow/control
21	14	Improve government/community leadership
22		More revenue
23	3	More industry
24	4	Improve/provide zoning (countywide)
25		Improve operation of government
26		Prevent abuse of welfare
27		Improve race relations
28	6	More paved streets and highways
29		More blacks in labor force
30		Remove railroad crossing downtown
31	17	More vocational training/facilities
32		Proportional representation on school board
33		Improve community pride/spirit
34		Extend city limits
35		More diversified economic base
36		Less political factionalism
37	32	Public kindergarten/day care
38	18	Revitalize downtown
39		Improve health programs
40	10	Improve/expand water system
	5	More/better medical facilities
	7	More doctors
	21	Drug use prevention program
	22	More stores/shopping facilities

Figure 7.3: Mean Rank Order of Community Needs — 1968–1975 (Continued)

Mean Rank Order		
4 Large Communities	4 Small Communities	Needs
	24	Dentist/dental clinic
	25	More/improved motels
	27	More interest and participation in religion
	28	Beautify city/county
	31	Sanitary landfill
	33	New public health building
	34	Higher wages/income
	35	Provide civic center
	36	Facilities for elderly
	37	Fire protection
	38	Enforce health regulations

tearing down old buildings as the community's problems (see Figure 6.5).

Social analysts, using quality of life indicators (see Chapter 3) have attempted to determine objectively what problems small towns may have that require national attention. They contrast large and small (metropolitan and nonmetropolitan) places and have generally found that in small places:[2,1]

(1) family income is lower;
(2) the proportion of poverty households is higher;
(3) the number of older aged persons is higher;
(4) housing is more likely to be dilapidated or without plumbing;
(5) there is slightly higher infant and maternal mortality rates;
(6) there is considerably less health care services;
(7) there is considerably lower crime rates, both violent and property;
(8) there is only marginally adequate public services;
(9) existing and potential sources of revenues are ineffectively used;
(10) governments lack the capacity to deal with changing economic, social, and political factors: weak and poorly defined authority, nonrepresentative councils, inadequate administrative and planning support services.

It should be noted that quality of life statements in themselves do not constitute problem statements. Local decision makers must extract problems from specific settings by analyzing their situation. Some proceed from an ideal set of goals and objectives, contrasting the reality of the local situation to assess whether it measures up to the preferred state of affairs. Some follow trends and calculate whether conditions are getting better or worse. Others contrast the situation in their town to that found elsewhere. Still others "know" what they want and need. All of these provide the basis of dissatisfactions in the sense that things are not as they ought to be and indicate potential problems. An actual problematic situation requires someone to say they prefer to reduce the discrepancy between what ought to be from what is. To do so generally requires some sense that there is a solution within reach, there are others who agree, and there is a willingness to take action.

People express a desire for change, but mainly within traditional norms and using traditional institutions. When pressed, however, on what kind of change and how it is to be achieved, the answers are not clear or shared. Those engaged in community development face this difficulty. Achieving agreement about what constitutes a community problem is no easy task nor are solutions easily found as to what should be done about the problem. The difficulty comes in part from how small towners view themselves as individuals with citizen responsibilities, what is known about the community, what goals are agreed upon, what resources are available, and what degree of willingness there is to do something about a recognized problem. Despite some broad areas of agreement when discussing community affairs at the store or over the dinner table, turning good intentions into action often exposes hidden levels of disagreement and dissention.

Rarely, if ever, does a problem exist in complete isolation. Each problem should be seen as part of a set of interrelated problems, and solutions to most problems may produce other problems and aggravate the total situation. Ackoff refers to the interrelated system of problems as a *mess*. He believes that efforts to break down messy problematic situations into simpler discrete problems "not only usually fails to solve the individual problems involved, but often intensifies the mess."[22]

How, then, are the solutions to community problems normally decided upon in the small town? Residents and decision makers, believing the small town to be a simple social system, assume that problems are only temporary disorders that can be readily solved through the use of common sense or intuition. If the problem is one of dilapidated housing, the solution is to tear the slum area down. If the problem is seen as unemployment, the solution is to attract into town an economic firm that will provide jobs. Thinking in these simple terms, coming up with intuitive answers to problems usually results in small towners treating the symptoms of problems, seldom delving deeper into the more fundamental causes and implications of the problem.

We believe the community analysis framework helps those engaged in community improvements to see some of the interrelationships. Take, for example, the desire to solve the community problem of a shrinking economic base which results in unemployment, closed retail stores, declining tax base, and so on. In response the small town is likely to undertake a number of actions without a concomitant understanding of their impact on the community. Industrial development sites or parks will be designated, the municipality will offer potential investors tax concessions and land cost "write-downs," roads and other infrastructure (sewage treatment, water lines, and the like) might be improved, and job training at the government's expense would receive strong consideration. Additional considerations *should* be given to the social, economic, political, and cultural implications. For example, consider the following possible effects of economic development on a community's four dimensions.

(1) Social Structure. Economic development brings with it an influx of new residents, who have to be housed. Factories are not easily hidden or ignored, and as such, combined with the location of new residences, alter the ecology of land use. The influx of new workers also might change the town's social class structure. Without understanding these impacts and how to deal with them, economic development may "solve" the problem of local unemployment only to generate other equally significant problems.

(2) The Local Economy. New economic firms may alter the town's economic base, its integration into the local economy, as well as the community's stability and autonomy. The rush to industrialize often overlooks the impact of development on the community's image as a small town, its desire to protect the environment, and the role of the public sector in supporting the investments of private, profit-making institutions.

(3) Influence Patterns. New economic firms generally affect the local power structure. An absentee-owned company will obviously look out for its own interests and thereby is likely to make the town more dependent on its decisions. Local decision makers now may have more important decisions to make and a need to respond to a whole new set of demands for tax concessions and public services.

(4) Cultural Norms. The decision whether to seek the establishment of a new economic firm should raise value-related questions pertaining to the preferences and consequences of growth and change. If they are not explored, they may reappear as future problems. Once industrial development is undertaken, urbanization generally follows and frequently the process is irreversible.

None of the early renewal efforts in Coalstream proceeded beyond recommending a physical revitalization. A regional religious organization came to town to do something about housing, but soon left after finding a lack of resources—money and leadership—to achieve its objectives. Two comprehensive plans produced a long list of recommended projects that could have been carried out through state and federal grants: (1) a million dollar sewage treatment; (2) a half-million dollar demolition project of blighted residential and commercial areas; (3) the construction of a new city hall; (4) improved park and recreation facilities; (5) construction of high-rise senior citizens housing; (6) a downtown parking garage; (7) construction of a new highway linking the town to a nearby larger city; (8) construction of a new "planned unit housing project development"; and (9) preparation of an urban renewal proposal. All these recommendations were presented as standard fare to not only Coalstream but to other small towns without regard to their feasibility or acceptability to local residents and leaders.

Community Profile of Conditions

It is important to construct a community profile on the prevailing conditions as part of the community development process. Finsterbusch and Wolf refer to profiling as providing a "baseline social data of the impact area from which the magnitude and intensity of changes, induced and incidental, can then be estimated.... The profile features on which data are gathered comprise the salient impact categories of late assessment."[23] The community analysis framework provides a useful format to gather information and facilitate an assessment of relevant conditions. We propose the use of the discussion guides in Chapters 2 through 5 which include the items found in Figure 7.4.[24]

The community profile can facilitate the search for items to be placed on an action agenda; that is, a careful reading of community needs as expressed by the rank and file citizen and leaders should be assessed against the community profile. A proposal to meet a community need must run the gamut of the prevailing value system and be acceptable to the leadership so that community support can be mobilized behind it and ample resources—financial and human—can be mustered.

We developed a community profile in Chapter 6 by applying the discussion guide to Coalstream. It is a statement on the conditions prevailing in 1974. There are, of course, some references to the past. Any effort to revitalize Coalstream would involve some contemplation of a future state of affairs. Valachos proposes that one can use the profile not only as a statement about contemporary conditions but to explore the three major conceptions of futures as he describes them.[25]

(1) *Potential* futures, based on varying degrees of likelihood ranging from probable through possible to plausible. Such futures indicate the range of reasonably expected events, to potential developments, even to more remotely imaginable futures.
(2) *Preferable* futures raise the question as to what are the desirable or normative ends and goals that we would like to attain or perhaps should pursue in a consciously planned society.
(3) *Practicable* futures involve essentially a "synthesis" between expectations and desires. In other words, practicable futures reflect the synergistic effect of joining the most probable paths suggested by present conditions (based on past trends) together with desired goals or targets.

Purposes of Community Projects

While it may seem that each project has a unique purpose, there are distinguishable patterns. Some goals center on the community while others center on individuals. Some objectives are manifest and openly articulated, while others are latent or unclear to the

Figure 7.4: Community Profile

Values (Chapter 2)
    Important decisions and/or issues
    Institutional roles
        – government
        – economic firms
        – schools
        – religions
        – family
    Ranking budget priorities
    Quality of life preferences

Social Structure (Chapter 3)
    Socioeconomic status
        – income
        – occupation
        – education
    Cultural factors of differentiation
        – race
        – ethnicity
        – neighborhood
        – religion
        – life-style
        – age
        – sex
    Geographic patterns
    Modes of interaction
        – coexistence
        – neglect
        – cooperation
        – competition
        – conflict

Local Economy (Chapter 4)
    Indicators of employment and sources of income
    Economic base
    Locational factors
    Economic function

Patterns of Influence and Decision-Making (Chapter 5)
    Electoral participation
    Attendance at meetings
    Governmental forms
    Prevailing influence patterns
    Community resources
    Major revenue sources of local government
    Expenditure patterns

community developers. All are a matter of value judgments and guide the actions of those that participate. Most project objectives become subject to heated discussions if not outright disagreements. Often the project is linked to the people who initiate, support, and advocate the change, both as to its substantive and procedural merits. Hytten, for example, emphatically believes that the values and norms of all the project participants, including the developer, should involve two conditions.[26]

> One is a higher level of explicitness as regards the ideological (or normative) component, i.e., a more conscious recognition of the fact that, whether we like it or not, our choices and actions in such a field as socio-economic development are ideologically charged. It is thus better to embrace one's norms and values openly than to be unconsciously guided by them. The other is a clear insight into the necessary relationship between values and fact, and particularly into the ways in which our value systems—whether conscious or not—influence our choices of the facts we consider relevant to a certain course of action.

The purpose of most community developments projects is, of course, focused on improving the community itself. There are three general objectives often cited—each seems to have a time frame: past, present, and future. Those that focus on the past prefer a previous state of affairs over that of the present as they perceive that things are getting worse. These proponents are generally nostaligic about the good old days. Their orientation is remedial rather than aspirational. They do not resist change but wish to undo previous changes. They frequently find themselves "swimming against the stream," facing considerable opposition. The once successful boom town that now is in decline is most likely to engage in projects that are designed to capture the essence of the past.

The second focus is to maintain the community essentially as it is. This preference for the status quo does not necessarily mean inaction, because some activity is generally required just to preserve the present set of conditions. To do nothing is to invite decline and obsolescence. After all, the physical plant—water lines, Main Street, and so on—wears out, if not properly and periodically maintained or replaced. To prevent decline and/or deterioration these projects attempt to preserve the present condition of the community with minor changes here or there, primarily so that it does not become dysfunctional.

The third focus is future-oriented and directed toward preparing the town for a set of alternative futures. Ackoff refers to this orientation as preactivism, whose advocates

> are not willing to settle for things as they are or once were. They believe that the future will be better than the present or the past, how much better depending on how well they get ready for it. Thus they attempt to predict and prepare. They want more than survival; they want to grow—to become better, larger, more affluent, more powerful, more many things. They want to do better than well enough; they want to do as well as possible, to optimize.[27]

They attempt to deal with problems before they arise and become serious. They seek to exploit opportunities and plan for the future.

Community development projects are also deliberately undertaken to enhance individual life-styles, or at least result in improving people's lives in a number of ways. One is to increase self-awareness, encourage the individual to aspire to achieve his or her full potential. Another is to enhance the citizen's role in collective decision-making and to develop the ability to articulate individual demands and assume leadership responsibilities in mobilizing community resources.[28] Still another develops the skills of individuals to enter the marketplace with useful talents which will enhance their employment.

The Coalstream Project involved a comprehensive assessment of the revitalization potential of the community through an open-ended probe of resident preferences for plans for the future. The explicit objectives were:

(1) to determine the nature and extent of diverse revitalization potentials and the interests and capacities of town residents in reaching and sustaining these potentials;
(2) to develop the framework of a program which could help realize potentials through concentrated, resident-based action and the judicious concentration of outside resources;
(3) to suggest a general model by which state, regional, and local-level agencies can determine the revitalization potential of many other small towns.

Types of Strategies for Community Development

Those who engage in community development universally emphasize the process aspect of their work. Saunders for example defines it as "the deliberate attempt by the community people to work together to guide the future of their communities, and the development of a corresponding set of techniques for assisting community people in such a process."[29] Schler has elaborated a general model with three dimensions—procedural, content, and human interaction—to promote locally planned change.[30]

Consideration should be given then to how to proceed with the process of community development. That is, what approach should be used to fulfill the needs of the community, and how does one implement the proposed agenda item of community action. Warren considers three major types of strategies to bring about purposive change.[31] Each corresponds to the "issue situations" as well as to the existing power configuration, the target population, and the timing of the proposal. He believes that the selected strategy for purposive change should be based on the community response to issues. That is, three issue situations can be analytically distinguished on a continuum.

The first is *issue consensus* where there is basic agreement on the way the issue is to be resolved or there is a likelihood of reaching agreement once the issue has been fully considered. Consensus may arise from common interests based on common values or a convergence of interests when values differ. The second is *issue difference* where there is yet no agreement that the proposal actually addresses an issue (the developer is trying to create an issue) or there are differences on the substance of the proposal itself. The chief contributor to the former difference may be apathy, to the latter, simply opposition. In either case the developer believes that through persuasion the opponents will perceive their "true" values or interests, and agreement will be reached. The third is *issue dissensus* where important parties refuse to recognize the issue or oppose the developer's proposal and there is little likelihood of achieving any consensus through substantive discussions.

In the consensus situation Warren would use a *collaborative* strategy. The role of the developer is that of enabler or catalyst. The objective is to help the relevant groups reach agreement. In this situation, differences are assumed to be based on misinformation or poor communications and the chief obstacle is inaction. Once the participants understand the deeper nature of the problem there may be little consensus. Those using the collaborative approach therefore adapt by either restricting the process to parties with similar value configurations or by narrowing the issues to those on which agreement can be reached. In the issue difference situation Warren would use a campaign strategy. The role of developers here is that of "persuaders." They seek "agreement through the use of persuasion, pressure, bargaining, and moderate coercing of consent." In the issue dissensus situations Warren would use a *contest* strategy which temporarily, at least, abandons efforts toward consensus and instead attempts to further one's own side of an issue dispute. The role of the developer is that of "contestant." Warren lists four types of contests:[32]

(1) vigorous clashes *within accepted social norms;*
(2) attempts to *change the distribution of power;*
(3) *violations of the usual community norms;*
(4) *conflict* in the deliberate attempt to harm opponents or remove them from the issue.

The contests may take place in each of four elements of our community analysis. In fact, Coleman has discerned that community disputes arise in four areas of life:[33] (1) economic, (2) power or authority, (3) cultural values or beliefs, and (4) persons or groups.

In the beginning the strategy used in the Coalstream Project was collaborative with the Borough Council as part of issue consensus over physical development. Campaign strategies were used when disagreements occurred over which housing options should be adopted.

Conclusion

As big cities continue to be in trouble, the small town appears to increase in importance to many Americans. The preference to live in small communities remains stable while that for big cities declines. People desire to be free from congestion, noise, crime, pollution, and the like while seeking greater opportunity for such leisure time activities as hunting, fishing, and gardening. Small towners emphasize the ambiance of the smaller community—a small place maximizes independence and familiarity. One's opinions and participation count more in small places. Thus the small town is perceived as a reasonable alternative to the big city.

There are some 33,000 municipalities and towns in America with less than 10,000 residents. Each town seems special to its residents and certainly there is a wide range of places whose uniqueness should be preserved if the residents prefer that it remain that way. Small towners, however, disagree on this matter because some want to bring the conveniences of an urban society to their community while others reject the forces that tend to lure them into the urban maelstrom.

Small communities not only have problems but they need help. Williams, however, questions those who have been helping.[34]

Planners who stamp out master plans for small communities which all look alike (under the spiral binder and impressive graphs) and which all too often read alike (down to the recommendation for an industrial park site and a new village hall). The idiosyncracies and generic traits of each small town are all but ignored.[35]

Engineers who charge stiff fees for water and sewerage feasibility studies (less than 1 in 10 of which is ever implemented according to an informal study of our Institute). More distressing are urban solutions (such as the package treatment plant for sewerage) scaled down and presumed to fit the small town. The idea of technologies appropriate to a smaller scale is rarely considered.[36]

Governmental and private agencies who continue to administer social "services" which are "delivered" to individual "clients." This entire approach should be questioned by small towns. It denies their capacity to act independently to deal with at least some problems of residents on a community basis.

Academics who study small communities with a framework analysis implicitly urban and who do not stay long enough to learn of important, if subtle, distinctiveness. Also, the scholars most often leave town to report in professional journals, seldom bothering to return to present information to those who might directly use it.

Perhaps part of the difficulty lies with the assumption of community developers. Brokenska and Hodge have found in the development literature a number of fallacies:[37]

(1) communities have a unity, and shared interest; cooperation is natural, even if it is not now present;
(2) where there is a strong sense of community, development will be easier, and have more immediate results;
(3) "tradition" generally inhibits progress, and development can free people from the traditional restraints;
(4) communities are good, and enshrine sacred values;
(5) the "felt needs" of a community exist and can be discovered by cross-examination;
(6) each community has a clearly defined leadership and if no leader appears at once, then a stranger must show the way;
(7) everyone desires a higher level of living and welcomes change;
(8) community development is not only more lasting (and cheaper) than direct action but also better as it builds self-reliant citizens.

Even planners have come in for critical comments. Cohen points to a number of failures on the part of planners to recognize the distinctiveness of the small town setting:[38]

(1) they often assume that some historical imperatives—natural economic forces occurring regardless of consciously directed human will—have created present conditions as a result of their own momentum;
(2) they compose an urban framework on the small community input because it is the only conceptual framework familiar and comfortable to the planners;
(3) they apply inappropriate scale in their solutions which is beyond the comprehension and capabilities of the residents and officials of the town;
(4) they have an obsession with economic theories that negate the need for the consideration of other factors in community development. More specifically they include Streeten's four fallacies of economic analysis:[39]

  (a) one central factor explains the need for intervention;
  (b) once certain conditions are fulfilled, other developments will automatically follow;
  (c) treat each community problem separately;
  (d) misplaced selection and measurements of variables central to small town development;

(5) they rely on standard solutions while ignoring the specific context of plans and decisions, thereby placing them in social and administrative isolation from perceiving the social and political ramifications of their solutions;

(6) they primarily consulted with community elites, where as those without power and influence are routinely and systematically overlooked.

We have presented a framework for community analysis that we believe should help avoid many of these fallacies. Furthermore, we believe that a sensitive application of this framework should include a careful reading of the degree of penetration by outside socioeconomic and political forces, the kind of needs and problems, and the shape of the community profile. In addition, the purpose of the project should be made explicit and the appropriate development strategy should be thoughtfully selected.

We would be remiss if we failed to state that this book is only a step in a more comprehensive process that should reconnect community action and theory. What is still needed is an action-research agenda that will provide the basis for meaningful generalizations. Activists and theoreticians should be working side by side so that they can more effectively share and use their insights and experiences to the advantage of small towns and small towners.

NOTES

1. Robert A. Dahl, *Who Governs* (New Haven: Yale University Press, 1961), pp. 180-183.

2. Robert Presthus, *Men at the Top* (New York: Oxford University Press, 1964), p. 95.

3. Dahl, op. cit., pp. 63-84.

4. William H. Form and Delbert C. Miller, *Industry, Labor and Community* (New York: Harper & Row, 1960), p. 47.

5. J.F. Steiner, *The American Community in Action* (New York: Holt, Rinehart & Winston, 1928).

6. James West, *Plainville* (New York: Columbia University Press, 1945).

7. Maurice R. Stein, *The Eclipse of Community* (Princeton: Princeton University Press, 1960).

8. Robert Lane, *Political Ideology* (New York: Free Press, 1962), p. 226.

9. John Kolb, *Emerging Rural Communities* (Madison: University of Wisconsin, 1959).

10. Don Martindale and R. Galen Hanson, *Small Town and the Nation* (Westport, CT: Greenwood, 1969), p. xiv.

11. Ibid., p. xiv.

Community Analysis and Considerations for Development

12. Joseph Bensman and Arthur J. Vidich, *The New American Society* (Chicago: Quadrangle Books, 1971), p. 346.
13. Roland L. Warren, *The Community in America* (Chicago: Rand Mcnally, 1972), p. 16.
14. Ibid., p. 19.
15. Ibid., p. 20.
16. U.S. Senate, Committee on Agriculture, Nutrition, and Forestry, Subcommittee on Rural Development, *Rural Development Oversight,* 95th Congress, First Session (Washington, D.C.: Government Printing Office, 1977), p. 23.
17. Martindale and Hanson, op. cit., p. 60.
18. Ibid., p. 60.
19. Bert E. Swanson, *The Concern for Community in Urban America* (New York: Odyssey Press, 1970), p. 133.
20. Harold L. Nix, George S. Brooks, and Bradley C. Courtenay, "Comparative Needs of Large and Small Communities," *Journal of Community Development and Society,* Vol. 7, No. 2, Fall 1976, p. 101.
21. Vary T. Coates, *Revitalization of Small Communities: Transportation Options* (U.S. Department of Transportation, Office of University Research, May 1974), pp. 9-21.
22. Russell L. Ackoff, *Redesigning the Future* (New York: John Wiley, 1974), p. 21.
23. Kurt Finsterbusch and C.P. Wolf, editors, "Profiling" in *Methodology of Social Impact Assessment* (Stroudsburg, PA: Dowden, Hutchinson & Ross, 1977), p. 153.
24. For another list of factors from which to draw a small community profile see U.S. Senate Committee on Agriculture and Forestry, *Small Community Needs,* June 1970, pp. 21-28.
25. Evans Valachos, "Assessment, Forecasting and Alternative Futures," *The Social and Economic Impacts of Highway Projects,* edited by Gordon Enk (Rensselaerville: Institute on Man and Science, 1976), p. 280.
26. Eyvind Hytten, "Is Social Development Possible?" *International Review of Community Development,* December 1969, p. 15.
27. Ackoff, op. cit., p. 25.
28. William W. Biddle and Lourlide J. Biddle, *The Community Development Process* (New York: Holt, Rinehart & Winston, 1966), pp. 157-158.
29. Irwin T. Sanders, "Theories of Community Development," *Rural Sociology,* Vol. 23, March 1958, pp. 4-5.
30. Daniel Schler, "The Community Development Process," *Community Development as a Process,* edited by Lee J. Carey (Columbia: University of Missouri Press, 1970), pp. 130-137.
31. This discussion draws heavily on Roland L. Warren, *Truth, Love and Social Change* (Chicago: Rand McNally, 1971), pp. 7-34.
32. Ibid., pp. 24-45.
33. James S. Coleman, *Community Conflict* (New York: Free Press, 1957), pp. 5-6.
34. Harold S. Williams, "Smallness and the Small Town," *Small Town,* Vol. 8, October 1977, p. 10.
35. Many of these plans were done with federal support—i.e., the so-called "701 program" of HUD. Given the availability of public money and the requirement that municipalities must have a plan to be eligible for certain kinds of assistance, the master plan proliferated in the 1960s. The proportion of plans that have been implemented—even carefully read—is abysmally miniscule.
36. One encouraging sign is a series of 1977 conferences sponsored by the Environmental Protection Agency on lower cost waste-water treatment approaches for the small

community. It should have happened a decade ago; but better late than never.

37. These have been drawn from David Brobenska and Peter Hodge, *Community Development* (San Francisco: Chandler, 1969), pp. 20-23.

38. Richard Cohen, "Small Town Revitalization Planning," *Journal of American Institute of Planners,* Volume 43, January 1977), pp. 4-9.

39. Paul Streetan, "The Use and Abuse of Models in Development Planning," in *The Teaching of Development Economics* edited by Kurt Martin and John Knapp (Chicago: Aldine Publishing, 1970), pp. 89-93.

GLOSSARY

AMENITIES: Pleasantries, conveniencies, and contributions toward comfort and aesthetics made in the community once baseline conditions for survival (e.g., water and sewage) are met.
ATTENUATED FAMILY: A family with only one parent and one or more children.
ATTITUDE: A disposition reflected in the general state of readiness to react to something rather consistently in a certain kind of favorable or unfavorable way.
AUTHORITY: A legally or formally vested power to make decisions and to enforce them through the use of sanctions (force) if necessary.
BALANCE OF TRADE: The relationship between payments owed for merchandise and services bought by a person, household, community, or other group and payments claimed for merchandise and services sold.
BELIEF: A predisposition to respond in a preferential way to the object of that belief. Beliefs may describe an object or situation as true or false, evaluate it as good or bad, or advocate it as desirable or undersirable. Beliefs involve a person's knowledge and feelings and when challenged, will often lead to some sort of action.
COERCION: The use of force or punishment, whether visibly or invisibly, to get someone to act or behave in a certain way.
COMMUNITY: A group of people sharing a limited physical area in which they carry out at least some important daily activities, such as eating, sleeping, child-rearing, participation in clubs and groups, and the like.
COMPETITION: The pursuit of goals by two or more people, groups, communities, or other entities under conditions such that the achievement of a goal by one group usually prevents its achievement by the others.
COMPLEMENTARY INDUSTRIES: The presence in a community or other setting of industries wherein the functioning of one enhances or makes possible the functioning of the other. Thus, the industry of a saw mill and the industry of making wooden toys from the product produced by that saw mill are complementary.

COMPROMISE: The settlement of a disagreement reached by mutual adjustments or concessions by those involved. Bargaining is often a method used for reaching a compromise—frequently used in communities to settle controversies.

CONFLICT: Behavior of two or more persons, groups, or communities in relation to each other such that an objective of each group is to deflect, change, or in other ways stop the other.

CONSENSUS: Opinion, thoughts, or feelings that are in general agreement among those involved. Consensus may be a baseline tradition (e.g., people just naturally agree) or may be consciously brought about in a planned way (e.g., broadening a decision made so that there is something in it for everybody and they will agree to it).

CONSUMPTION: The use of goods and services. Consumption and its end product, waste, are important means of measuring the total economic activity of an individual, family, or community.

CONTROVERSY: Arguments or disagreements among individuals or organizations who differ in their assumptions about something, about a problem or issue, or what should be done about it. Controversy is generally thought to be stronger than mere disagreement and yet not as strong as conflict.

COOPERATION: Pursuit of goals by two or more people, groups, or communities under conditions where achieving the goal by one person or group helps the others also to achieve that goal.

CULTURE: The concepts, habits, skills, institutions, and so on of a given people in a given period which result in certain patterns of behavior and material products; similar to civilization.

DECISION-MAKING: The process of defining issues, problems, and/or options to be taken in such a way that the decision between or among alternatives is made; decision-making in communities is often an ongoing process of interaction involving both official leaders and many residents.

DEMOCRACY: The opportunity for the members of a given unit (e.g., a society or community) to participate freely and with effect in the decisions which affect their lives, and the ability of the system to respond to their collective preferences. The term often includes both direct democracy in which individuals represent their own interests and representative democracy in which members of the society elect certain people to represent their interests in government.

DISTRIBUTION (ECONOMIC): The process and end result of getting goods and services to consumers. One aspect is the marketing and transportation of goods and services to people. Another is how those goods and services get apportioned among the members of a community or society.

DIVERSIFIED: Made up of many rather than few kinds of elements. Thus, a diversified economy in a small town is one made up of different kinds of activities and not just one or two businesses or industries.

DOCTRINE: A law or principle accepted by a group of people who believe in it as a philosophy or school of thought. Thus, the concepts that small towns are either valuable to our future or a thing of the past are both doctrines which are associated with different people who believe in them.

DYNAMICS: The pattern of change, growth, evolution, decline, or other form of change on an individual, community, or societal level. Many things which are thought to be static (e.g., unchanging) when viewed from a distance are in reality dynamic, even in the most stable of small towns.

ECONOMIES OF SCALE: The principle that the average cost per unit of output produced declines as the size of an organization increases until, at some point, the unit cost increases. Thus, the more cars, televisions, or hamburgers a firm sells, the lower, presumably, the cost of each unit. Recent interest in smallness, whether small business, small communities, or small tools, suggests that there are often corresponding economies of small scale which industrialization has tended to hide.

ELITES: A term used to define or to indicate those few individuals at the "top of the heap"—such as those who with a higher status (such as based on social standing, income, education) or those who get to make or enforce most decisions (e.g., the political elites). The notion of elite implies that even in the most democratic of situations there are usually those who have more of whatever is valued than do others.

ENTREPRENEUR: A person who initiates and develops new organizations or institutions under conditions of high risk.

EXTERNALITIES: The by-products or spill-over effects, whether desired or not, of activities which affect others. Thus, the pollution experienced downstream caused by the manufacturing plant dumping wastes into the water upstream is an externality affecting residents downstream.

EXTRACTIVE INDUSTRY: Industries which obtain products directly from the land or water, including farming, lumbering, mining, quarrying, and fishing. It is important to differentiate in extractive industries which take renewable resources (such as plants and lumber) and those which extract nonrenewable or irreplaceable resources (such as minerals).

FEDERAL SYSTEM: A form of government in which authority is shared between a central overall administration and state, regional, or other forms of local government.

HETEROGENEOUS: Characterized by people, groups, or other elements who differ fundamentally or in kind. Thus, residents of a small town with different backgrounds, values, opinions, or vocations would form a heterogeneous community.

HINTERLAND: Applied to regions and areas lying beyond the boundaries of cities or metropolitan areas, yet which relate to those urban places in a systematic, ongoing way.

HOMOGENEOUS: The quality of being of a similar kind or nature.

HUMAN ECOLOGY: The patterns of how people settle and use the land and develop social and cultural processes upon it.

IDEOLOGY: Beliefs, attitudes, and opinions such that a coherent and regular pattern of thinking in an individual or group emerges. Ideology is usually applied to more fundamental and important aspects of life. Thus, one might be said to have an ideology governing political events or human relationships, but an opinion about preferences for one breakfast cereal over another.

IMAGE: A mental picture or map of a place based upon what an individual brings to bear when he "looks." Thus, memories, associations with other things, and preconceptions all play a role in one's image of a small town.

INFLUENCE: The ability of a person, group, or other organization to convince another to behave in a way that the first party prefers or desires and in a way the other would not have behaved necessarily.

INFRASTRUCTURE: The underlying foundation of physical (roads, sewage, water) and social (education, and so on) services upon which the economic and social development of a community or nation depends.

INSTITUTION: A significant practice, relationship, or organization in a society or a community, generally valued by the society for more than just the services or benefits institutions provide; usually, institutions are connected to the stability of societal or community functions.

INFLUENTIAL: A term used by social scientists to designate an individual who is believed to play a significant role in decision-making. Influentials are those people who over the course of time tend to get their way more than their share of the time.

LEADERSHIP: That role in a group or community referring to the coordination and/or concentration of influence exerted to make decisions and take planned change actions. Leadership can be concentrated (e.g., one leader) or it can be shared (e.g., many people provide leadership in areas in which they have special interest or knowledge).

LEGITIMACY: The belief of a group of residents or citizens that a government, authority, or institution has the right to rule and that it ought to obey the rules and laws of that entity.

MOBILIZATION: The condition of involving and energizing a group of citizens to take action on a certain issue or problem in a concentrated organized fashion.

MULTIPLIER: The way in which an increase or decrease in the strength of one factor causes effects in others. For example, if an increase in one manufacturing job stimulates the creation of three service jobs, then the multiplier is three.

MUNICIPAL BUDGET: The community's formal written estimate of income and expenses for a future period, usually one year. The budget provides important information on the sources of local government income and, just as important, on the priorities for allocating public funds as determined in the political process.

NONPARTISAN: Refers to the lack of formal political party labels and, ideally, influences. Thus, a nonpartisan election is one in which candidates do not run as Republicans, Democrats, or even independents.

NORM: A shared understanding in a community or group of what is correct and incorrect behavior in many common given situations. Norms are often built upon shared beliefs and values.

NUCLEAR FAMILY: The basic unit of family organization, which includes both parents and their children by birth or adoption. (The nuclear family is thus differentiated from the extended family which can include aunts, uncles, grandparents, and others who are biologically related, yet not "nuclear.")

POLICY: The intended purposes, mechanisms, and guidelines by which programs are carried out in groups, communities, and the nation. Policies are usually long-range commitments for which more immediate programs can vary greatly.

POLITICS: Those processes of human action by which decisions affecting many people get made. Political issues almost always involve distinction and tension between the "common good" and the interests of groups and individuals at a more private level. Politics involves the use of or struggle for influence or power, often defined in terms of control over the vital resources needed to get one's way in a given setting.

POWER: The extent to which a person or group can impose its will upon others. Power is evident in a small town in any situation when one group or person gets others to do what it wants.

PRAGMATIC: Believing or acting in terms of practical consequences of belief or action. The pragmatist is concerned with a sense of what will work, often with no regard for theory or even consistency.

PRESTIGE: The respect or honor enjoyed by individuals or groups in a community or other setting on the basis of their property, lineage, occupation, education, income, influence, or whatever else might be admired within a community.

PRODUCTION: The process of creating goods and services through the translation of resources (money, raw materials, labor) into products (goods and services) which are desired and/or needed.

PROGRESSIVE TAX: Tax whose rate increases as income increases. Thus, a progressive tax might assess a person with an income of $5,000 at 10% levy (e.g., $500), but assess a person with an income of $10,000, $2,000 (e.g., 20%). A progressive tax is associated with the idea of income redistribution—helping to redistribute money from those who have more through a higher tax to those who have less through a lower one.

PUBLIC INTEREST: Concern for the general welfare of all the people in a community, state, or nation. Although many people talk of the public interest, it is very difficult to pin down and, in many instances, may not exist at all. That is, different people have such different ideas of the public interest that what in reality we have is many different private interests, each of which compete for the primacy of their expression.

QUALITY OF LIFE: The quality of the human experience as judged both by environmental factors (e.g., degree of pollution, overcrowding, housing, adequacy of health care, and so on) and personality factors (e.g., sense of happiness, frustration, disappointments, fulfillments, and so on).

"RANK AND FILE": Those below the level of top leadership in a given group or organization, including business, the community, the church, or virtually any other institution.

REFORM: A concept of change involving the modification or elimination of elements of a situation or system considered highly undesirable. Thus, a reform platform for a politician is often associated with eliminating corruption and increasing public participation in decision-making.

"RULES OF THE GAME": The ways in which members of a community are expected to behave; how they are to make demands, process them into policies, and influence their implementation.

RULING CLIQUE: A small group which controls community decision-making with formal and/or informal authority. The notion of a clique suggests a small group of like-minded people who join together for a specific purpose of gaining leverage in a community to make decisions.

SANCTIONS: Rewards which are promised for compliance or punishments promised for noncompliance to given guidelines, commands, or expectations. Thus, breaking a law warrants the negative sanctions of fines or jail, while performing well in one's job may bring bonuses and pay raises.

SELF-INTEREST: Concern and support for those things that directly affect a person whether positively or negatively. Self-interest is not necessarily selfish interest in that it can be shared by many people and be a positive force in decision-making that will benefit everyone. Yet, the adding up of all self-interest in the community may well not total to the greatest degree of public interest.

STATUS: A position based upon prestige and life-style and accorded to individuals and groups in the community or other social settings. Status can come from wealth, income, or occupation, but may also be the result of family background, the use of power, admired talent, or friendship with others of high status.

STRATIFICATION: A system whereby those of different status are spread out along a continuum of high status to low status based upon recognizable "strata" into which different people fall. Many social scientists believe that there is a stratification system in all communities, at least to some extent, and the expression "We're just plain folks here" really isn't true beneath the surface.

STRUCTURAL UNEMPLOYMENT: The long-term unemployment of a sector of the local population due to an insufficiently flexible supply or use of the area's potentially productive resources. It may be that capital and material resources are not being used to their highest, most profitable efficiency and thus businesses leave many employable workers unable to find a job. Or, workers may remain unemployed either because they are not trained to work in other industries nearby or because they are unable and unwilling to move to another location. In any event, it is the basic structure of the community and the setting which has caused unemployment. This differs from short-term unemployment (for example, irregular plant layoffs or closings where workers are quickly rehired elsewhere), or cyclical unemployment (periodic inadequate demand for workers), such as is the case with the seasonal demand or lack of demand for farm help.

SUBSIDIARY COMPANY: A company controlled by another company such that the controlling company owns 50% or more of the shares of the subsidiary company or through some other mechanism directs and controls the subsidiary. Many once-independent industries in small towns have been purchased in the past decades by larger corporations and are now operated as subsidiaries. This arrangement is said to create external linkages between the community and the world beyond it in that the major decisions in the economic sphere are then made by individuals and groups, such as corporate executives, outside the community.

VALUE: A type of belief expressing the degree of worth and importance in an area of behavior and existence deemed important by a person or group of persons. Values can be abstract ideals, such as truth and beauty, or much more specific preferences, such as security and risk-taking. A value system involves an ordering of values in terms of their importance and specifies the interrelationships between values.

VOLUNTARY ORGANIZATION: Those organizations which people join willingly and in which they are involved without formal remuneration. Some are institutionalized, such as the Rotary Club or PTA, and others are ad hoc groups formed around specific issues.

AUTHOR INDEX

Ackoff, Russell L., 253, 259
Adams, Bert N., 122
Agger, Robert E., 50, 62, 190
Aiken, Michael, 169-170
Andrews, Richard B., 125, 128
Arensberg, Conrad M., 57
Beale, Calvin, 18-19, 136
Bell, Colin, 35
Bensman, Joseph, 28, 55-56, 62, 87-88, 108, 125, 178, 243
Berelson, Bernard, 46, 161
Bernard, Jessie, 92, 96-97, 101
Berry, Brian J.L., 127
Bert, Kendall, 126-127
Beshers, James M., 102-103
Biddle, Loureide J., 31-32, 66, 259
Biddle, William W., 31-32, 66, 259
Blalock, H.M., 108-109
Blizek, William L., 65
Blumenfeld, Hans, 129
Bonjean, Charles M., 85
Brinkman, George L., 128, 136, 137, 140, 141
Brobenska, David, 263
Brooks, George S., 250-252
Bruyn, Servenyn T., 110
Burgess, E.W., 91
Campbell, Angus, 49
Campbell, Donald T., 47
Carter, Steve, 126-127
Cary, Lee J., 31
Cederbloom, Jerry, 65
Chapin, F. Stuart, 127
Clark, Terry N., 104, 165-166, 172
Clavel, Pierre, 144
Coates, Vary T., 26-27, 252
Coleman, James S., 64, 261
Converse, Phillip E., 49
Courtenay, Bradley C., 250-252
Dahl, Robert A., 170-171, 239
Dahrendorf, Ralf, 83
Darwent, D.F., 143
Davis, Allison, 87, 88-89

Davis, Kingsley, 19
DeTocqueville, Alexis, 43
Doeksen, Gerald A., 131, 132, 135-136
Donohue, George, 64
Duncan, Otis D., 84-85, 131
Durr, Fred, 126
Elazar, Daniel J., 61-62
Enk, Gordon A., 48
Finsterbusch, Kurt, 255-257
Form, William H., 59, 124, 131, 240
Free, Lloyd, 20, 49-50
Freire, Paulo, 35
Fugitt, Glenn, 18-19, 22
Fuller, T.E., 123
Gagnon, John H., 18, 122
Gallup Poll, 20, 47
Gamson, William A., 102
Gardner, Burleigh B., 87, 88-89
Gardner, Mary R., 87, 88-89
Gibbs, Jack, 19
Gilbert, C.W., 169
Gingrich, N.B., 123
Gold, Raymond L., 48
Goldrich, Daniel, 50, 62, 190
Goldsmith, William W., 144
Goodall, Brian, 126, 127
Gordon, Milton M., 89
Goudy, Willis, 137, 139
Gross, Bertram, 85, 113
Hamilton, Richard F., 99-100
Hanson, R. Galen, 57, 243, 248
Harris, Chauncey O., 92, 130
Harris, Louis, 49
Heady, Earl O., 121
Hill, Richard J., 85
Hillery, George A., 161
Hodge, Peter, 263
Hollingshead, August B., 28
Hornick, William F., 48
Hoyt, Homer, 91-92
Hytten, Eyvind, 258
Jacobs, Philip F., 46
Jansma, J.D., 123

Johansen, Harley, 22
Jones, Charles, C., 112
Khinduka, S.K., 32-33
Kimball, Solon T., 57
Kolb, John, 242, 264
Kuehn, John, 131, 132, 135-136
Lipset, Seymour M., 44
Ladd, Everett Carll, 50, 62
Lane, Robert, 50-51, 242
Lenski, Gerhard, 58
Linton, Kenneth, 24, 51-54
Long, Norton E., 102
Lunt, Paul S., 86-87, 95
Lynch, Kevin, 94-95, 103
Martindale, Don, 57, 243, 248
Marx, Karl, 82-83
Mathur, O.P., 127-128
McKenzie, R.D., 91
McLemore, S. Dale, 85
Milbreath, Lester, 180
Miller, Delbert C., 59, 124, 131, 240
Miller, S.M., 85, 98-99, 113
Mishan, E.J., 122
Morse, R.M., 127-128
Mott, Paul E., 131
Murphy, Michael J., 123
Myrdal, Gunnar, 83-84, 96
Newby, Howard, 35
Nie, Norman H., 162-163, 180-181
Nix, Harold, 250-252
Norbert, Peter, 126-127
Ogburn, William F., 131
Park, Robert E., 91
Pope, Liston, 58
Poplin, Dennis E., 124
Poston, Richard W., 41
Presthus, Robert, 37-38, 55, 62, 167, 178, 179, 239
Quandt, Jean B., 107-108
Redfield, Robert, 35
Rein, Martin, 85, 113
Reissman, Leonard, 83, 90
Richards, Robert O., 137, 139
Richardson, Harry W., 141, 142-143
Roby, Pamela, 85, 98-99, 113
Rodgers, William L., 49
Rogers, David L., 137, 139
Root, Brenda, 138
Sanders, Irwin T., 125, 259

Schler, Daniel, 260
Schnore, Leo F., 101
Schmidt, Joseph, 131, 132, 136, 135-136
Schulze, Robert O., 167
Schumacher, F.F., 17, 122
Schwartz, Richard D., 47
Sechrest, Lee, 47
Simon, William, 18, 122
Smith, Page, 131
Smith, Robert T., 130
Sofranko, Andrew, 138
Srole, Leo, 96
Stein, Maurice R., 242
Steiner, Gary A., 46, 161
Steiner, J.F., 242
Stewart, Guy S., 65-66
Strauss, Anselm L., 96
Streeten, Paul, 263
Summers, Gene F., 139, 140
Svalastoga, Kaare, 82
Swamy, M.C.K., 127-128
Swanson, Bert E., 24, 25, 29, 42, 50, 51-54, 62, 169, 170, 178, 190, 248-249
Swanson, Edith, 42, 178
Thomas, R. Murray, 100-101
Thompson, Wilbur, 129, 133-134, 136, 143
Tuan, Yi-Fu, 45
Tweeten, Luther, 128, 136, 137, 141
Ullman, Edward L., 92
Valachos, Evans, 257
Verba, Sidney, 162-163, 180-181
Vidich, Arthur, 28, 55-56, 62, 87-88, 108, 125, 178, 243
Warner, W. Floyd, 28, 86-87, 95, 96
Warren, Roland, 41, 44-45, 61, 161, 243-244, 260-261
Watts, William, 20, 49-50
Webb, James, 47
Weber, Max, 82-83, 157
West, James, 242
Wildavsky, Aaron, 168, 178
Williams, Harold S., 17, 24, 51-54, 262
Williams, James, 138
Williams, Robin M. Jr., 89-96
Wolf, C.P., 255-257
Zimmerman, Joseph F., 174

SUBJECT INDEX

Authority, 158
Business firms, 135, 231
Central business district, 123
Charlestown Village, 54-55
Citizen participation, 45, 156, 181
City manager, 159
Community
    action
        categorical, 33-34
        comprehensive planning, 34
        dialogical, 34-35
        mobilization, 172-173
    analysis, 35-38, 109, 179, 237-241, 254-255, 264
    as economic mechanism, 124
    controversy, conflict, 46, 62-64, 102, 104-109, 210-212
    cooperation and competition, 101-102, 145, 210-212
    decision-making centers, 157-160
    decision-making organizations, 110
    decisions and issues, 67-68, 155, 160, 179, 217-218, 226-230
    developers, 27-28, 30-31, 32-33, 64-65
    development, 64-66, 109-110
        fallacies, 263-264
        purpose (ends), 257-259
        strategies (means), 260-261
    improvement, 64
    mobilization, 172-173
    needs and problems, 249-255
    organization, 161
    priorities, 72-75, 189, 232-234
    profile, 191, 255-257
    renewal, 110, 179, 192
    resources, 170-172, 185-187, 220-223
    revitalization, 123, 259
    studies, 28
    types of
        absorbed, 248-249
        coopted, 247-248
        ignored, 245-246
        self-reliant, 246
        transitional, 247
    values and norms, 77-79, 226
Decentralize, 122
Democracy, 156, 178
Discriminatory behavior, 108-109
Economic base, 128-129, 136, 138, 148-150, 194-195, 199
Economic
    "boosters," 137-138, 144, 192
    functions, 129-133, 150, 198-199
    multiplier, 139-140, 144
    political, 239-240
    progress, 136-317
    revitalization, 137, 143
Economies of size, 141
Education, 113, 201
Employment, 123-124, 139-140, 192-195, 207
Equal opportunity, 81
Equality, 44
Equity, 144-145
Ethclass, 89
Ethnic group, 89, 203
Family, 57-58, 72, 207, 232
Futures, 63, 75-77, 235-236, 257, 259
Governance, 22-23, 212
Government, 59-60, 70-71, 157, 182, 223-224
    scope of, 62, 155, 173-176, 231-232

Government spending, 49-50, 177-178, 188-189, 225-226
Growth points, 141-143, 145
Helvetia, 53-54
Human ecology, 91-93, 103-104
Ideologies, 50, 55-56, 62
Income, 97-99, 111-112, 143, 147-148, 194, 200-201, 207
Industrialization, 137-141, 143, 242
Inequalities, 81-82
Influence, 157, 164, 212, 215-220
Inner core, 241
Institute on Man and Science, 35, 51, 236
Intergroup contact, 89-90
Labor force, 146-147, 196
Labor market analysis, 133-134, 143
Land use, 91-95
Leadership, 101, 164-170, 179, 183-185, 215-220
Life styles, 85, 87, 93, 103, 114, 205-206, 259
Local economy, 59, 124, 145
Local politics, 156
Locational analysis, 126-128, 135, 138, 145, 148, 196-198
Mayor, 158-159
Mobilization of bias, 99-101
Neighborhoods, 90-95, 114-117, 204, 207-210
Occupation, 85, 112-113, 134, 201
Participant observer, 48
Penetration of small towns, 242-249
Planners, 93-94
Policy makers, 25-27, 30-31
Political culture
  individualistic, 61
  moralistic, 61
  traditionalistic, 61-62
Political participation, 99-100, 108, 161-164, 213-215
Political-social, 241
Population size, 18-20, 39, 86, 141, 143, 158, 169, 174, 179, 239
Power, 157
Power elite, 110
Poverty, 22, 143, 144, 207
Public opinion, 47

Quality of life, 97-99, 252
  preferences, 75-77
Reformers, 107, 160
Regionalism, 133, 140-143, 145, 199
Religion, 58-59, 72, 204-205, 231
Revitalization, 17
  normative neutrality, 65-66
Rural, 19-20, 121
  ghettoization, 135-136
Schools, 56-57, 71-72, 229-230, 231
Settlement pattern, 135
Sex-age, 114, 206-207
Smallness, 17, 122
Small town
  contrast to big cities, 24, 48-51, 170
  death of, 18, 122
  decline, 135-137
  defined, 19
  problems, 21-24, 160
  ways of looking at, 29, 35
Smalltowners, 24
Social caste, 83-84, 88-89
Social class, 81-83, 108, 110, 144
  ideologies, 87-88
Social-economic, 240-241
Social interaction, 95-97
  modes, 101-102, 108, 117, 210-212
Social justice, 66
Social mobility, 106, 108, 178
Social scientists, 28-29, 46, 91, 125, 157, 167
Social space, 90-95, 114-116, 207-210
Social stratifications, 82-85, 102-109, 113-114, 117
  cleavages, 108
  mechanism of, 84-85
Socioeconomic status, 85-90, 99-109, 133, 200-203, 212
Stump Creek, 51-53
Taxes, 97-98, 137, 140, 143, 155, 176-177, 187, 197-198, 218, 224-225
Urban, 19-20
  "urban-size ratchet", 136, 141
Urbanization, 21-22, 122, 242
Values institutions, 56-60, 69-72
Voting, 163-164, 180-181, 212-213

ABOUT THE AUTHORS

Bert E. Swanson is Professor of Political Science and Urban Studies at the University of Florida. He received his Ph.D. from the University of Oregon. He has had extensive experience in governmental service at the national (aide to a U.S. senator), state (aide to a State Commissioner of Education), and local (aide to a mayor, the Board of Education, and the Public Housing Authority) levels. Professor Swanson has engaged in urban research throughout America and in the revitalization of small towns with The Institute on Man and Science. He has authored and co-authored numerous books and articles on government and community studies which include: *The Rulers and the Ruled* (Woodrow Wilson Award of the American Political Science Association), *The Struggle for Equality, The Concern for Community in Urban America, Black-Jewish Relations in New York City,* and *Discovering the Community.*

Richard A. Cohen heads Richard A. Cohen Associates. At the time of writing this book he was Senior Planner at The Institute on Man and Science. He received his Master's degree from the University of Pennsylvania where he was awarded the Charles Abrams Scholarship of the American Institute of Planners. He has engaged in consultation for governmental officials and agencies and conducted studies in poverty and urban neighborhoods. He was Editor of *Planning Comment* and has written numerous articles on small town revitalization.

Edith P. Swanson is Director of the Florida Center for Community Analysis. She received her Master's degree from Sarah Lawrence

College. Her experience includes community action and extensive research on public education, mental health, and other public policy areas both at the University of Florida and in a consortium with Cornell University. She has coauthored a number of articles and books including *Discovering the Community*.